Climate Change Impacts on Plant Biomass Growth

Mohammad Ali

Climate Change Impacts on Plant Biomass Growth

Springer

Dr. Mohammad Ali
Department of Environmental Science
  and Management
North South University
Dhaka
Bangladesh

ISBN 978-94-007-5369-3   ISBN 978-94-007-5370-9 (eBook)
DOI 10.1007/978-94-007-5370-9
Springer Dordrecht Heidelberg New York London

Library of Congress Control Number: 2012947986

Printed on acid-free paper

Springer is part of Springer Science+Business Media (www.springer.com)

Contents

Illustration of Symbols

C_i	Carbon content of the continuous input where $i = 1$, f, b, s, w stands for leaf, fine root, branch, stem and woody roots respectively.
C_m	Continuous input of soil organic matter through the decomposition of litters.
D_p	Lifetime of wood productions.
$g_c(t)$	Quantity of stored carbon in time (t).
g/g/day	Gram per gram per day.
g_m	Carbon content in old growth tree.
Gt	Gega tonne.
h_b, h_s, h_w	Discrete input of branch, stem and woody root material that is left over after harvest.
H_i	Discrete input with $i = 1$, f, b, s, w.
k_i	Fractional rate of decomposition of litters to soil organic matter where, $i = 1$, f, b, s, w.
k_m	Fractional rate of decomposition of soil organic matter to gaseous component.
LAP	Leaf Area Partitioning.
L_1, L_f, L_b, L_s, L_w	Continuous litter accumulation from leaf, fine root, branch, stem and woody root respectively. If suffixed by (t) it means the accumulation material within that time of rotation.
Mg C ha^{-1}	Million grams of carbon per hectare.
Mkm^2	Million square miles.
MPa	Mega Pascal.
Mt	Million tons.
μ_i	Fraction of litter compartment that becomes soil.
μl/l	Micro Litre per litre, equiv. to ppm.
NAR	Net Assimilation Rate.
NPK	Nitrogen, Phosphorus and Potassium.
NPP	Net Primary Production.
Pg	Petagram $= 10^{15}$ g $=$ One billion tons.

ppb	Parts per billion.
ppm	Parts per million.
P_x	Different types of wood products, where $x = 1, 2, \ldots ..x$.
RGP	Root Mass Partitioning.
T_m	Maximum mean annual increment.
T	Time (rotation time).
Tg	Tetra gram.
U_j	Fraction of removed material used for product j.
$Y_{(t)}$	Fractional quantity of stored carbon in each gram of tree material.

Chapter 1
Introduction

Plant biomass plays an important role in sustaining environment and life including human life on earth. The earth system is a closed system except for its absorption of solar energy. Green biomass is the only component of the earth system that captures bursts of solar energy, converts them, and allocates them to flows to other components and consumers of the ecosystem. In an ecosystem, where green biomass is the primary producer, humans occupy almost all positions, primary, secondary, or tertiary, within the consumer chain. Humans derive their essentials – food, fiber, fuel, and fodder – directly from green biomass and indirectly through other primary and secondary consumers of green plants. Hence, performance of green biomass growth in an ecosystem is not only important for smooth functioning of all components of the earth system but also for human existence in a befitting manner.

Besides obtaining energy from it, humans use the ecosystem environment for recreation, sport, aesthetics, shelter, clothing, medicine, and education. For these, the human being tends to regulate the environmental factors involved in the growth and development of green biomass. However, climate change, the unpredictable and irreversible change in environmental conditions, poses serious risks and uncertainties as to how the vegetation will respond to these changes, firstly because the changes are complex and slow to reveal their influence in a generation and secondly they are irreversible. Due to the climate change situation, the success of human beings in maintaining a sustainable environment will largely depend on the art and the science of using the changed environmental factors for green biomass growth.

The art and science, that is, the management, of environmental factors requires interplay of a number of "P" words – application of *pre*-acquired knowledge of *pro*fessionals in a *p*lanned way so that *pro*curement of *pre*dicted *pe*rformance and *pro*fit are *po*ssible from a *pro*duction system. In this regard, management of green plant biomass occupies a special situation. The management of green biomass, as professional forest managers adopt their appropriate roles, usually involves application of several environmental regimes for manipulation of growth of seedlings in order to produce mature trees within a particular ecosystem – a process

M. Ali, *Climate Change Impacts on Plant Biomass Growth*,
DOI 10.1007/978-94-007-5370-9_1,
© Springer Science+Business Media Dordrecht 2013

that may take several decades to come to fruition. Many things may happen within this long growing period – including affliction of an unpredictable effect of climate change on the growth of seedlings and their final utilization.

There are lower green biomasses, like green bacteria, phytoplankton, and green fungus, that can have several life cycles within the long periods required for higher green biomass growth, for example, tree growth. Therefore, climate change-induced effects on a lower green biomass could have differential impacts at different stages of a higher green biomass growth. As a consequence, in formulating management prescriptions for vegetation, forest managers must think about complexities of factors that could have potential influences on future growth and should take as many necessary actions as possible today to minimize their impacts tomorrow. Otherwise, society will have to pay a great and inimitable future price for the neglect/errors committed today.

Thus, in addition to predicting the demand for and supply of forest products, forest managers must pay close attention to predictions of periodical environmental changes, the services of which are most likely to be in great demand in the future. The managers have to understand that there could be possible substitutes for forest products; however, there will be no substitute for services that a forest can provide to nature through its green biomass. Perhaps, we, the human beings, have to risk our lives, if services from green biomass are reduced under changed environmental conditions of nature. The pity does not end there; we have to remember that changes in environment are cumulative, multifarious, irreversible, and borderless and are likely to bring further changes in site, climatic factors, nature, and impact of diseases. All these changes are likely to have snowballing effects on future growth and performance of green biomass.

Due to changes in the climate situation, if there is lack of understanding in these unpredictable changes and if appropriate actions are not taken in appropriate time, the multifarious changes between environment and vegetation are likely to continue endlessly with increasing amplitude until a final end is struck – "the dooms day." Vegetation, as a primary producer in the ecosystem and as sustainer of all other components of the ecosystem, will compel other components of nature to oscillate in similar irreversible fashion causing loss of many of the existing biodiversities and allowing a few rapidly evolving species to survive and dominate the environment. The nature of such dominance of a few is likely to shatter the present dominance of human beings.

The survival of the few will bring a complex change in growth, form, volume, or height based on differential physiological processes under different climate change environments, which, in their turn, will bring detrimental changes in sustainability conditions. There is no way that human beings can deal with such complexities unless the green biomass is taken care of appropriately. We have to understand clearly that when we say we have to manage environment sustainably, we can do so only through manipulating the performance of biotic components of our ecosystem. Human beings have no capability to control physical environment like rainfall, sunlight, and atmospheric circulation. The interaction between green biomass and physical environment is mutual. Because green biomass is linked with all other biotic and abiotic components of our ecosystem, taking care of vegetation can bring changes to other components of the environment as we desire. Performance of green biomass remains the central point of all kinds of sustainability that we need.

Under natural conditions, the influences of environmental factors on physiological processes of green biomass can be expressed through both the nature and the distribution of growth of plants to their different components. Growth of plants and their specific organs depend on their age, environment, and genetic makeup. Hence, in using the term "growth," we should consider the plant's growth not only in volume or height but also its distribution, reproductive potential, branching, root development, and other apparent responses. We might even discuss the future evolutionary development of plants in terms of their exposure to long-term environmental factors under the climate change situation. There are many environmental factors that can influence the processes involved in the growth of green biomass. The three major environmental influences are:

1. Those associated with the lithosphere involving root processes
2. The biosphere involving questions of competition and association
3. The atmosphere involving gaseous and light processes

These environmental processes are intricately related and very difficult to discuss in an independent way. In this book, we will take only one issue of climate change environment at a time – and then will try to explain how it may likely affect the growth of green biomass. The objective is that, if we can configure the changes in green biomass, it would be possible to anticipate the subsequent consequences on primary and secondary consumers. Further, although the consequences of climate change impacts on green biomass growth could be complex and need coining with technical terms, we wish to keep our presentation as simple as possible so that there remains food for all audiences including general readers, students, academicians, researchers, modelers, and scientists alike. To begin with, let us see what climate change means to us.

Chapter 2
Climate Change: What Is in the Name

Climate change is a very well-discussed phenomenon. Still, we appear to be shocked when we hear about it. By climate change, we understand detrimental effects in environment – emission of pollutants, temperature rise, precipitation change, sea level rise, flooding, intensified cyclones, abrupt frequency of events, ozone layer depletion, biodiversity loss, vegetation change, and drought – almost all negative impacts. Indeed, climate change is an umbrella concept. Scientific evidence that humans were changing the climate first emerged in the international public arena in 1979 at the First World Climate Conference (WCC) (Depledge and Lamb 2005). Since then, it did not take long to reveal the dreadful consequences of climate change. By 1988, when IPCC was formed, the dangerous consequences of climate change became clearer.

The root causes of climate change remain in the human-induced change – global warming. However, most discussions on climate change stem from its politics rather than its physics. The physics of climate change remain somewhere in the greenhouse effect problems of the lower atmosphere. It is imperative that we understand about the greenhouse effect and global warming and their linkages with climate change to understand the climate change impacts on green biomass and to plan for mitigation. We will take up those issues in a separate chapter. Here, we will give a synthesis on climate change concepts – to clarify what exactly the name climate change involves.

Climate, as we understand, is the average behavior of weather of a large tract of land over a period of time. Therefore, unlike "weather," which means temperature, rainfall, humidity of a particular time at a local area, "climate" has spatial and temporal components. It consists of many kinds of weather events, their periodicities, intensities, and nature of dynamism. However, when we talk about climate change, we actually understand more than the literal changes of weather events over the time and space of that climate. For example, if we take the ozone (O_3) layer depletion, it is not actually a weather change but is included among harmful impacts of climate change. Similarly, changes in ocean current, melting of ice, and loss of biodiversity are included in climate change; they are not discussed as conventional events of weather. Therefore, the motivation of deriving and using

M. Ali, *Climate Change Impacts on Plant Biomass Growth*,
DOI 10.1007/978-94-007-5370-9_2,
© Springer Science+Business Media Dordrecht 2013

Fig. 2.1 Climate change dimension

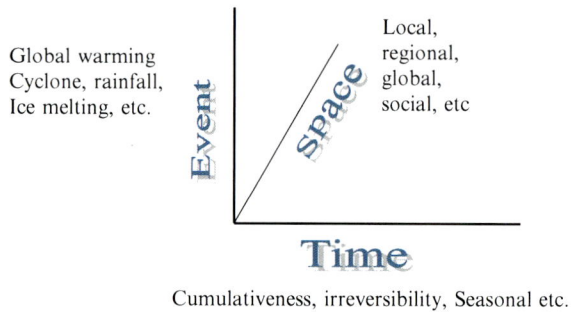

a term "climate change impact" is mainly to include broader effects of changes in the events than that of elements of weather. The events, of course, are the results of average behavior of elements of weather.

We have to say a few words here to clarify for readers what we mean by dimensions of climate change. When we say climate change impacts, we do not refer to the impacts of weather events only; we also include their aftermath over time, their relations with other components of the ecosystem, and especially their impact on human beings. This means that the term climate change has a spatial as well as a temporal boundary. The spatial boundary of climate change determines its scale, whether local, regional, or global, and the temporal boundary explains the nature of accumulation and reversibility of such changes over time.

Figure 2.1 gives a schematic view of different dimensions of climate change with some examples from each dimension. When we say climate change, we understand all of the dimensions – not just the events. When we say "weather event," that event might be either useful or harmful for human beings. However, when we consider those events under climate change, our emotions induces a negative connation. It is like pollution. Pollution is the increase of those ingredients that are harmful for human beings. So, increase of CO_2 is pollution, but increase of O_2 in the atmosphere is not considered pollution because it is useful. The concept of climate change also exerts a similar pressure on our emotions. With this philosophical meaning in mind, we present here a brief description of each of the dimensions of climate change.

2.1   Temporal Dimension

Temporal dimension indicates the vertical arrangement of environmental events and accumulation of their impacts over time. On the time scale, if we take the geological time, for example, Triassic, Jurassic, and Cretaceous, we find each of the scale considers millions of years at a time and have some major events associated with it (Table 2.1). These geological events are end results of climate events on the time scale. If we look at the events individually, we can see that the end results have

Table 2.1 Major environmental and evolutionary events in the geological records

Geological eras		Age million years	Major environmental events
Quaternary		0–1.64	Greenhouse effect; ice age
Tertiary	Pliocene	1.64–5.2	Early hominids
	Miocene	5.2–23.3	North and South America join; ancestor of honeycreepers arrives on the Hawaiian chain
	Oligocene	23.3–35.4	Evolution of grasslands and grazing animal
	Eocene	35.4–56.5	India collides with Laurasia
	Paleocene	56.5–65.0	Rise of mammals; extinction of the dinosaurs
Mesozoic	Cretaceous	65.0–145.6	Rise of dominance of angiosperms; breakup of Gondwanaland
	Jurassic	145.6–208.0	Early angiosperms; early birds; breakup of Pangea starts
	Triassic	208–245	Early mammals
Paleozoic	Permian	245–290	Mass extinction event
	Carboniferous	290.0–362.5	Huge rain forests of tree ferns, club mosses, and early gymnosperms
	Devonian	362.5–408.5	Early terrestrial vertebrates; first seed plants
	Silurian	408.5–439	First land plants
	Ordovician	439–510	Metazoans
	Cambrian	510–570	Rise of metazoans; first fossil exoskeletons
Pre-Cambrian		570–	Origin of metazoans; origin of life

Adapted from Encyclopedia Britannica (2005)

not been repeated, which means that the changes that had happened over time were irreversible. The continuation of events has been prolonged over millions of years but not repeated. We may, with our modern scientific tools, find it possible to bring an extinct creature back, let us say from a remnant of DNA; however, there is no way we can bring the environment of that time back, even if we want to.

The greenhouse effect has been predicted as a major event in the Quaternary era. That event may continue for thousands of years with concomitant other environmental events. Those changes could be so complex and devastating that they will be unstoppable and unbearable. We, as human beings, may extend the temporal horizon through appropriate actions; however, if we do not take action in time, the consequences could come upon us sooner. We will be on the verge of extinction. Because we are the most helpless, we are the most vulnerable. Let us give some examples: ducklings can swim immediately after hatching from their egg; a calf can run immediately after birth. They have an instinct for survival. We do not. A human baby needs at least a decade and lots of training to become self-dependent. This means we use tools to protect ourselves. If we do not apply the tools in time and judiciously, they will not be useful for us. Similarly, temporal dimensions of climate change will likely appear as killers to our generation if do not take measures appropriately. The speed of evolution cannot catch the speed of climate-related devastation. Adaptation is a far remote situation without strong mutual cooperation among and across societies. Our instinct says, we cannot adapt without help.

The present climate change is happening very fast. It is not even 40 years since the world has detected changes in climate; by now, almost all countries are feeling the changes through fluctuation of climate events. Europe has seen many floods over the last few years; North America has witnessed a large number of tornadoes and category 5 hurricanes, unusual forest fires, and insect attacks on forest crops. Human-induced changes make the temporal dimension of changes very short. What used to be thousands of years for a change is happening in centuries, even in decades. Williams (2002) stated that the changes in the climate of the modern era began in 1750 along with the industrial revolution. The temporal speed of climate change is so high that if we do not take care of it now, perhaps we can never do it.

2.2   Spatial Dimension

The spatial dimension of climate change covers horizontal expansion. Horizontal dimension indicates the geographical coverage and distribution of events. On the spatial scale, we can take climate change as a phenomenon at global, regional, or local level. For example, changes in CO_2 concentration are happening globally; however, the intensity of weather change is different locally. We cannot segregate the issues of climate change in order to determine with certainty whether something happening in one region is or is not effecting some changes in another region. These changes do, but in a scale, and speed beyond our prediction. The understanding of horizontal dimension is important; taking actions by one or two countries, even if at the right time, will not be appropriate. The actions must be taken in concert.

For ensuring timely actions, perhaps we have to predict the overall magnitude, direction, and speed of climate change impacts; however, for concerted action, we may have to divide the concept of climate into its smaller components – thereby, we have to look at the extent and speed of variables that are changing on local, regional, and global scale to assure that they are in the right proportions and equitable ways, imposing individual and shared efforts through cooperation, collaboration, and assistance, whichever is possible under specific circumstances.

2.3   Event Dimension

The event dimension tells us the number, types, frequency, and intensity of calamities or changes that are likely to happen. If we look at the history, we see that the number of climatic events is increasing – for example, too hot, too cold, high rain, no rain, storm, cyclone, tornado, floods, vegetation change, biodiversity loss, sea level rise, and melting of ice. We have to classify and reclassify them because their intensity is changing; the frequency of occurrence is increasing; all of these changes are making climate change eventful. The more events we see, the more hopeless we have to be that we humans are doing enough to prevent climate change. However, these parameters of events give us indications about the nature of climate change.

2.4 Umbrella Concept

From the discussion on dimensions of climate change, we understand that there are relationships among the greenhouse effect, global warming, and climate change; however, they do not necessarily mean the same thing. Changes in temperature or weather events are due to a greenhouse effect of emission (gases that cause a thick layer in the troposphere) processes called "global warming." Global warming is a result of "radiative forcing effects" of a thick layer of greenhouse gases, in which long heat waves get reflected back to the earth, as a result of which the temperature of earth is gradually increasing. Due to the changes in temperature, the dynamics of other weather elements are also changing – which, along with other weather events, is a vital component of climate change. Therefore, when we describe climate change impact, mostly we will cover the impacts of changes of weather elements on vegetation.

The explanations also show that the influences of climate change are multifarious, mutual, and cumulative. This means that the very root cause of climate change – emissions – even if stopped, will continue their impact for a long time in different dimensions. Therefore, emissions could have both direct and indirect consequences on vegetation growth. As a result, when we wish to describe climate change impact on vegetation, we cannot avoid direct impacts of increase of emissions in the atmosphere. To be able to do so, we have defined climate change as an umbrella concept that includes all the phenomena of global warming and emission problems. Figure 2.2 presents the umbrella concept of climate change.

The figure shows that emission is not directly linked to the climate change umbrella; however, it is the prime reason for climate change influenced through

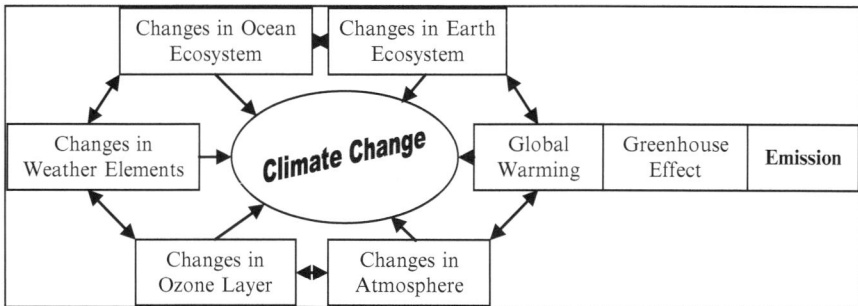

Fig. 2.2 Umbrella concept of climate change

global warming. One of the most important components of the emission problem is CO_2. It is also directly related to vegetation growth. The direct effects of emission on vegetation growth come mostly from an increase in the concentration of CO_2 in the atmosphere. Justus von Leibig (1803–1873), a famous German Chemist, first demonstrated that plants obtain their carbon from the atmosphere for photosynthesis and growth. Eventually, scientists, on the basis of absorbed carbon dioxide, have demonstrated many complex processes and have explained how plants maintain photosynthesis and growth. When the CO_2 concentration increases in the atmosphere, the nature and scale of these physiological activities may change and in this way eventually affect the nature of growth of plants. Since CO_2 is considered mainly responsible for climate change, and also has direct effect on plant growth, many examples will be presented in this book to explain the influences of CO_2 on plant growth.

Indirect impact of climate change on biomass growth may result from associated changes in the climatic variables caused by changes shown around the climate change in Fig. 2.2. The indirect effects of greenhouse gases are very complex. They have become critical issues because scientists are still not certain about their many components. Research is being conducted around the world to determine the nature of the paradigm "climate change" and to establish its possible socioeconomic and environmental consequences with other components of Fig. 2.2. The research effort is being extended from analysis of simpler thermometric data to those derived from satellite information strata, from lithosphere to stratosphere, and from local to global level. Green biomass is an important component of the biosphere as well as earth and ocean systems, having a vital role in socioeconomic and environmental homoeostasis processes. People occupy a key position within those systems, creating as well as controlling the processes; hence, consideration of human beings is given to reviewing the impact of climate change on the growth of plant biomass.

While the growth of plants is influenced by climate change effects, growing plants at the same time also influence the development of climate change. Plants take CO_2 from the atmosphere and use it in photosynthesis; plants also release CO_2 to the atmosphere through the process of respiration. The balance between the two is stored and appears as growth. Hence, growth in green biomass is itself a sink as well as a source for CO_2 within the atmosphere. Plant biomass may remain in a living form or in a dead form as litter and as part of the sequestered CO_2 entering the carbon (C) cycle. In this book, there will be a discussion explaining how some of the growth is dissipating through the C cycle.

Forest managers, the people associated with the management of vegetation and its growth, harvest some of the growth for different uses. The amount of growth harvested depends on the growth characteristics of the trees. Therefore, managers should appreciate all positive effects on growth, understand what may cause those effects, and consider all measures, which might be taken to best utilize those effects. At the same time, they also would like to avoid negative impacts. There may be one or more solutions among them that are acceptable. They could be a matter of preventing negative effects and/or ameliorating the positive effects, but the selection of appropriate measures should be aimed at supplementing the global effort and should be in harmony with regional conditions. Therefore, it is expected that the book will be useful for professionals dealing with forest and timber management.

One aspect of managed biomass growth is the maintenance of genetic diversity, not just species diversity, that is, the genetic variation within the species as well as the breeding of high yielding species. However, the greenhouse effect and the projected rates of climatic change could create complexities for the breeding program. Hence, in addition to reviewing some of the causes of greenhouse effect and its probable direct and indirect impacts on vegetation, it is important to discuss the impact on genetic diversity. Though genetic diversity is likely to be closely related to the regional climate, it will not be possible in this short essay to review the impact of the greenhouse effect on a regional scale. Based on the perspective of present scientific information, only certain aspects of climatic change are reviewed in general with some regional effects in particular. This has been done because climatically marginal countries like Bangladesh and the Maldives are more vulnerable to impacts of climate change. For these countries, whatever may be the enhancement of vegetation growth due to climate change would be meaningless if their very existence is threatened. An attempt is made to determine the options open to Bangladesh in combating such an insidious problem by considering its climate, soils, hydrology, and population, as well as its vegetation. Thus, the objectives of this study are to characterize and define the information available from other CO_2 research and to efficiently use that information to show how an increase in CO_2 concentration would affect the plant biomass growth.

There are some gases in the atmosphere and water vapor that have high radiative forcing ability. An excessive accumulation of these gases, especially CO_2 and CH_4, in the atmosphere produce a layer of sheath at the lower atmosphere through which longer heat waves cannot escape but are reflected back to the earth causing increased temperature of lower atmosphere. This radiative forcing action is analogous to the greenhouse effect adopted in controlled plant culture and, therefore, is known as greenhouse effect. Global warming is a cumulative result of greenhouse effects of these gases. Although the physical bases of climate change depend on greenhouse effects and global warming, climate change is a wider concept that merits clarification for understanding the specifications.

References

Depledge J, Lamb R (2005) Caring for climate: a guide to the climate change convention and the Kyoto protocol. Climate Change Secretariat (UNFCCC), Bonn, 27 pp

Williams M (2002) Climate change information kit. The United Nations Environment Programme and the Climate Change Secretariat (UNFCCC) and sponsored by UNEP, the UN Development Programme, the UN Department of Economic and Social Affairs, the UN Institute for Training and Research, the World Meteorological Organization, the World Health Organization, and the UNFCCC: Geneva, Switzerland

Chapter 3
The Greenhouse Effect

In 1986, Swedish chemist Svante Arrhenius built the theoretical foundation of what has become known as the "greenhouse effect." With remarkable prescience, he argued that if the carbon dioxide (CO_2) levels in the earth's atmosphere were to increase to double the present average level, the average temperature could increase by around 4–6 degrees centigrade (°C) with critical implications for the earth's climate (Falk and Brownlow 1989). Indeed, industrialization and escalating global populations over the last two centuries have placed unprecedented pressure upon the natural cycle of the earth. Agricultural and industrial activities have given rise to changes in the chemistry of the lower atmosphere as a result of gaseous emissions in quantities beyond the capacity of the earth's recycling systems. Certain gases in the atmosphere have a warming effect on the earth's surface, due to their ability to absorb heat being radiated back out to space – "radiative forcing" as it is called by Reid (1990). The irreversible effect of temperature increase due to this "radiative forcing" action of certain gases in the atmosphere is popularly known as the "greenhouse effect."

Scientists have identified carbon dioxide (CO_2), nitrogen oxides (NOx), methane (CH_4), ozone (O_3), chlorofluorocarbon (CFC-11 and CFC-12), and water vapor (H_2O) as the main greenhouse gases that are responsible for global warming. Among them, apart from global warming, some gases like CO_2, H_2O, and O_3 have direct consequences on vegetation growth by influencing physiological activities, and some others like CFCs have consequences on the outer ozone layer, resulting in less protection to survival of vegetation biomass. Therefore, though the greenhouse gas emission seems to be an atmospheric process, it actually influences the lithosphere as well as the biosphere simultaneously and in a vicious way through global warming. Thus, before discussing the impacts of climate change on growth processes of plant biomass, it is necessary to identify how the climate change events are related with the emission problem.

The "greenhouse effect" involves the processes and characteristics of emissions of certain atmospheric gases that are causing a change in temperature of lower atmosphere. "Global warming" is the end result of the greenhouse effect. Later on,

M. Ali, *Climate Change Impacts on Plant Biomass Growth*,
DOI 10.1007/978-94-007-5370-9_3,
© Springer Science+Business Media Dordrecht 2013

subsequent changes in climate and microclimate due to these temperature increases are also included under the terminology "climate change." Those changes could influence rainfall, sea level, salinity level, erosion, soil temperature, and microbial action. Rastetter et al. (1991) suggest that the greenhouse effect will result in a rate of increase of global mean temperature during the next century of about 0.3°C per decade. If it happens so, among other things, the resultant changes in atmospheric circulation and the partial pressure of gases in the atmosphere may affect adversely the normal physiological processes of vegetation and, hence, morphological and phylogenetic attributes of growth. As we all know, partial pressure of atmospheric gases plays an important role in physiological processes of plants.

3.1 The Greenhouse Gases

Since the beginning of the industrial revolution, human activity has significantly altered biogeochemical cycling at local, regional, and global scales. For example, Rastetter et al. (1991) estimated that over the years 1960–1990 (30 years), the pool of carbon (C) in the atmosphere (in the form of CO_2) has increased 670 pg (10^{15} g) C as a result of fossil fuel burning and forest clearing. Not that only human beings are exhaustive of resources, the effect of an increase in atmospheric CO_2 and other greenhouse gases due to their actions is causing great concern for the energy balance of the earth system. The main greenhouse gases and their possible sources are shown in Table 3.1. This table also shows the rate of change of concentration of these gases in the atmosphere.

Table 3.1 shows that the rate of increase of certain greenhouse gases, like CO_2 and CH_4, are more alarmingly increasing in recent periods (1990–2005) than those of the historical period. If we look at the sources of increase, we find that the causes related to human actions are mainly responsible except for some ruminants in the case of CH_4. Some other gases like CFCs were never present in the history before industrial development. They are new additions to the environment and human industrial activities are completely responsible for them. They were found extremely harmful for the ozone (O_3) layer, reducing its thickness in the stratosphere and causing solar UV rays passing through the ozone holes created through such dissipation of O_3 layer. The harm was more prominent in the polar region than in the equatorial region. Coincidentally, the countries in the polar region are developed. They have created a global pressure. CFC uses have been found that have been reflected in the reduction of use of CFC-11 as shown in Table 3.1. However, for other gases, such consensus for controlling has not been developed. When the cases of other gases like CO_2 arise, some countries are more careful about the health of their economy than that of humanity. As a result, emissions have been increasing in a soaring proportion. Climate change, thereby, we see that could be controllable by human beings is getting out of hand due to economic competition and political domination of states in the era of corporate colonization.

Table 3.1 Greenhouse gases, their sources and concentration in the atmosphere

Green house gas	Preindustrial conc. (1750)	Concentration (1990)	Trend % per year	Conc. 2005 (IPCC)	Trend % per year 1990–2005	Possible sources of increases
CO_2	275 ppmv	350 ppmv	0.4	379 ppmv	8.29	Fossil fuel combustion, deforestation
CH_4	750 ppbv	1,700 ppbv	0.8	1,774 ppbv	4.35	Paddies, ruminants, biomass burning, gas and coal fields, land fills, tundra
CFC-11	Nil	250 ppbv	4.0	251 ppbv	0.4	Industrial and consumer goods
CFC-12	Nil	450 ppbv	4.0	538 ppbv	19.56	Industrial and consumer goods
N_2O	285 ppbv	310 ppbv	0.3	319 ppbv	6.13	Fossil fuel combustion, biomass burning, agriculture
O_3 (trop)	15–20 ppbv	20–30 ppbv	0.5*	–	–	Urban and industrial pollution

Source: Perman GI (1996) and IPCC data

*Estimated to be 1 % in the Northern Hemisphere and 0 % in the Southern Hemisphere

3.2   Impact of Greenhouse Gases

We have stated that the impact of these gases on temperature of the lower atmosphere is a function of "radiative forcing" as well as their residence time in the atmosphere. Table 3.2 shows the effects and the relative properties of some of these gases. The table shows that among the known gases, CO_2 is the least effective gas. Nitrogen oxides are about 350 times and CFCs are enormously 15,000–27,500 times more effective than CO_2, considering their lower ranges of effectiveness. However, their concentration and hence the quantity in the atmosphere vary with CO_2 as the highest. In fact, the impact of greenhouse gas concentration in the atmosphere can be direct or indirect through the changes of climate and other environmental factors. Let us describe only a few of the gases and their characteristics in the atmosphere, which are important for biomass growth.

3.2.1   Carbon Dioxide

Carbon dioxide (CO_2) is the most abundant among the greenhouse gases in the atmosphere. It is the only one among the designated greenhouse gases available in ppm level (Table 3.1). Percent-wise, it is within 0.035 % of the atmospheric composition of tropospheric gases. In practice if it goes to 0.04 % of the atmosphere, what would be the turbulence in the climate change phenomenon is unimaginable. This gas has received most attention in the debate of greenhouse effects over the past two decades. The statement from the Toronto conference on the changing atmosphere, convened in June 1988 by the Canadian Government, which was the first comprehensive meeting between the climate change specialists and high-level policy makers of different countries, suggested to the governments that the emission of CO_2 should be reduced by 20 % by early next century to control the greenhouse effect (Reid 1990). But from Table 3.2, one can see that other greenhouse gases with very high radiative capacity and great atmospheric longevity may be collectively just

Table 3.2 Comparative impact of greenhouse gases

Greenhouse gas	Warming effect per mole relative to CO_2	Comparative residence time in troposphere	Relative greenhouse effectiveness (lower limit) compared to CO_2
C1	C2	$C3$	C4 = (C2 × C3)/60 (rounded)
CO_2	1	60	1
CH_4	36–70	10	6
N_2O	140–210	150	350
O_3	430–1,800	0.2	1
CFC-11	12,000–14,600	75	15,000
CFC-12	15,000–17,000	110	27,500

Source: Adapted from Reid (1990)

Table 3.3 Sources of carbon

(A) Atmospheric burden (10^6 tons as C)	
(B) Sinks and accumulation. (10^6 tons as C/year)	
Reaction with (OH)	490–1,150
Soil uptake	110
Accumulation (5.5 % per year)	10
Total	610–1,270
(C) Sources (10^6 tons as C/year)	
Fossil fuel combustion	190
Oxidation of anthropogenic hydrocarbons	40
Wood used as fuel	20
Ocean	20
Oxidation of CH_4	260
Forest wild fires (temperate zone)	10
Agricultural burning (temperate zone)	10
Oxidation of natural hydrocarbons (temperate zone)	100
Burning of savanna and agricultural land (tropics)	100
Forest clearing (tropics)	100
Oxidation of natural hydrocarbon (tropics)	150
Total	1,000
(D) Tropical contribution (10^6 tons as C/year)	
Tropical wetlands and savannas	80
Tropical biomass burning	20
Total	100

Source: McElroy and Wofsy (1988)

as important as CO_2. For example, if we look at the contribution to the greenhouse effect (Table 3.2), other gases in combination may contribute more than CO_2 to the greenhouse effect. Yet there has not been enough attention given to other gases.

Carbon in the atmosphere can be present as CO (carbon monoxide) or as CO_2. In fact, they change the state depending on the availability of free oxygen in the atmosphere. The global budget for CO_2 has been studied well; however, the estimate of quantity is less attended. Table 3.3 presents an estimate of the sources of C.

Table 3.3 shows that a major source of C is fossil fuel and oxidation of methane. Since methane is largely anthropocentric in origin, it may be possible that human activity plays a dominant role in the cycle of atmospheric C. Concentration of CO varies from locality to locality. For example, concentrations of CO as large as 200 ppb (parts per billion) were observed over the rain forest of the Amazon, with levels in excess of 400 ppb over a seasonally dry area (cerrado) (McElroy and Wofsy 1988). These differences are perhaps because O_2, given out by dense vegetation of the Amazon during photosynthetic process, is used by CO to transform it into CO_2. On the other hand, in cerrado areas, where density of green vegetation is less, CO is less likely to be converted to CO_2 other than whatever is available from the atmosphere. From there we may understand that increases in CO_2 concentration may likely have impact on concentration of O_2 in the atmosphere up to the limit of balancing of partial pressure.

If we consider that CO in the presence of oxygen (O) changes to CO_2, then much of the rise in atmospheric CO_2, for example, 315 ppm in 1958 to 345 ppm in 1976, 350 in 1990 (Table 3.1), could result in lowering the oxygen concentration. Emission of most CO is attributed to emissions from the burning of fossil fuel (Table 3.3). Nevertheless, the burning of vegetation could also have a significant role as a source for CO. But there are difficulties in estimating the fraction of CO released during burning. However, worldwide tropical forest burning only accounts for around 5 % of global carbon emissions and is dwarfed as a source of greenhouse gases by industrial emissions in the developed world (Cleary 1991). Yet, an amazing attempt is being made to reduce C emission through C trading through raising and conserving forests and REDD method. Little attempts have been made to reduce industrial emissions.

The massive emission of CO due to burning of fossil fuels is usually associated with large inputs of NO or NO_2 and reactive hydrocarbons in the atmosphere, providing an index mix for generation of photochemical smog. The concentration of oxides of nitrogen between 60 and 80 ppb may cause the elevation of levels of O_3 in the atmosphere. McElroy and Wofsy (1988) stated that the O_3 level usually declines above the forest canopy; this indicates a tendency of O_3 reaction with natural vegetation or with natural olefins emanating from the forest. So, there is a possibility that the tropical ecosystem will suffer damage by deposition of excessive ozone and other associated phytotoxic components of air pollution.

3.2.2   Methane

Methane (CH_4) is the second most abundant greenhouse "trace gas" (Table 3.1). Concentration of methane gas was increasing at a rate of about 0.8–1.0 % per year in both Northern and Southern Hemispheres (Dixon 1990, also Table 3.1). The rate of increase was more than double than that of CO_2. Information on long-term trends obtained from ice sheets of Greenland and Antarctica is provided in Fig. 3.1. The graph shows that prior to 1700, CH_4 maintained a concentration of approximately 0.70 ppm (parts per million), thereafter rising to 1.6 ppm around 1987 (also Table 3.1).

If we compare Figs. 3.1 and 3.2, we see that the increase correlates remarkably well with the growth in world population and indicates that the causes of CH_4 emission are most likely related to human activities, primarily agriculture. The sources of CH_4, as presented in Table 3.4, suggest a significant level of uncertainty about the production rate, which means that either too little research has been done on it or there are too many variables affecting the yield of CH_4. However, if we look at Table 3.2, we may see that CH_4 has 36 times more radiative power and in total is 6 times more effective than CO_2 in terms of climate change impacts. In that way, CH_4 emission is more significant than CO_2.

Table 3.4 also shows that animals are one of the major sources of methane. For example, small creatures such as termites, which are abundant in the forests,

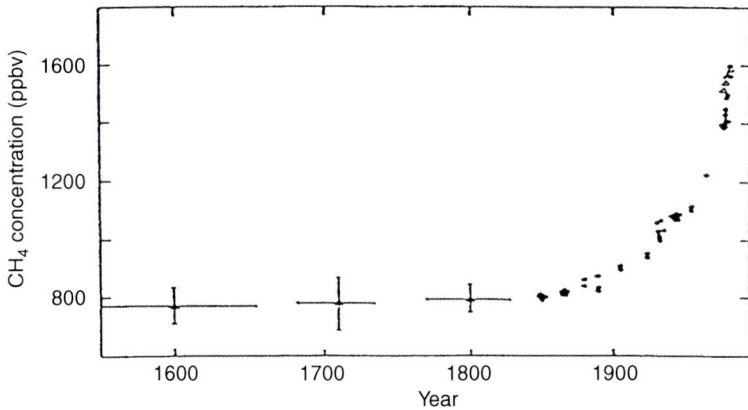

Fig. 3.1 Variations in CH$_4$ concentration over the past 500 years (Source: Dixon 1990)

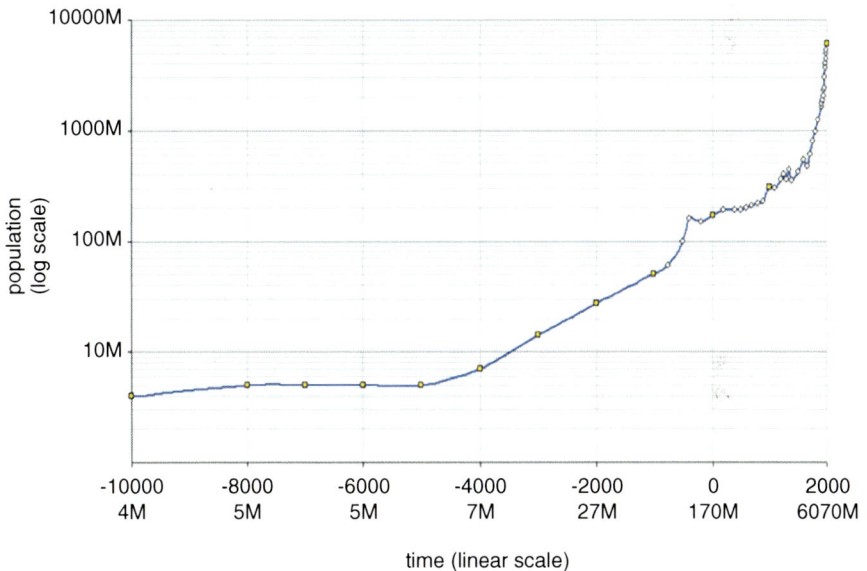

Fig. 3.2 Trend of global population increase in log scale (Source: Wikipedia)

have microorganisms in their gut causing emission of CH$_4$. The emission of CH$_4$ from the gut is highly variable, being a function of species, diet, and environmental conditions, and can range from 0.067 to 12 mg CH$_4$/kg termite/h which is equivalent to 1–20 mg/g of carbon consumed (Williams 1990). Williams cites the estimate of global emission of CH$_4$ from termites, based on the extent of certain ecological categories. From Australia alone, termite CH$_4$ production was estimated to be 0.80 Mt/y (million tons per year) from an area of 7.63 Mkm^2 (million square kilometer).

Table 3.4 Major sources of methane

Source	Production rate (CH_4/year) million tons
Ruminant animals	70–100
Insects	10–30
Termites	10–50
Rice paddies	70–170
Swamps, lakes	20–70
Tundra	40–110
Biomass burning	20–110
Landfill	30–60
Natural gas losses	20–50
Coal mining	12–40

Source: Dixon (1990)

CH_4 is also emitted from swamps and marshy areas. The major wetland groups are forested and non-forested swamps and alluvial areas. The two types of bogs which mainly occur between the latitudes 50° and 70°N are estimated to produce 56 % of the global wetland emissions (111 Mt/y), and the swamps which are largely situated in the tropical regions are responsible for 40 % (Dixon 1990). The total global emission of CH_4 comes from many sources such as mining, landfill, natural gas grid, motor vehicles, animals' ruminants, other animals, animal waste, rice field, biomass burning, termites, and wetlands. The Australian contribution of CH_4 could be around 4.7–15.7 (Mt)/y (Williams 1990).

Total atmospheric throughput of CH_4 is about $375 \pm 105 \times 10^6$ ($10^6 = 1$ million) tons of C per year out of which about 320×10^6 tons of C are removed by reaction of CH_4 with hydroxyl (OH) group in the atmosphere; about 5×10^6 tons of C is biological uptake, and the rest (about 45×10^6 tons of C) accumulates annually in the atmosphere. The sources of CH_4 emission are difficult to quantify. Production of methane mainly occurs from the C-rich system under anaerobic conditions through the action of fermentive bacteria. This means that C sequestrated in wetlands might come back in the form of CH_4 through the actions of lower organisms.

Table 3.5 shows that the agricultural activities contribute about 50 % of the total emission, with an additional 20 % from fossil fuels and from the burning of biomass. It appears that the tropics account for about one fourth of the current global source. Methane concentration in the atmosphere was assumed to be constant before AD 1600. Methane appears to have increased by more than a factor of 2.2 since the industrial revolution, more than 70 % of which is attributed to human activity. Direct measurement of CH_4 emission is difficult.

From Table 3.2, we see that CH_4 is at least six times more effective in global warming than that of CO_2. Methane's effectiveness in terms of the greenhouse phenomenon is related to its strong absorption band in the infrared, around 10 μ in wavelength, which is transparent to CO_2/H_2O absorption spectrum. The spectrum is near the wavelength of the peak in the outgoing earth radiation (Williams 1990). The incoming wavelength of sunlight is usually in the visible range of 1–10 μ; therefore,

Table 3.5 Sources of
methane emission

(A) Atmospheric burden ($3,500 \times 10^6$ tons as C/year)	
Reaction with (OH)	320
Uptake by dry soil	40
Accumulation (20 ppb/year)	45
Total	405
(B) Sources(10^6 tons as C/year)	13
Ocean	12
Tundra	25
Biomass burning	40
Natural gas loss + coal mining	95
Rice paddies	120
Cattle	50
Termites	50
Wetland + minor sources (by difference)	50
Total	405
(C) Tropical contribution (10^6 tons as C/year)	
Biomass	0.6
Soil erosion	7.4
Total	8.0

Source: McElroy and Wofsy (1988)

it mostly passes through atmosphere to the earth. However, the outgoing radiation from the earth is in the invisible infrared form and within the range of 10–100 μ. They are not transparent through the most atmospheric gases and are absorbed in the atmosphere.

3.2.3   Oxides of Nitrogen

Oxides of nitrogen are formed as by-products of combustion, either at the expense of nitrogen in fuel or by fixation of atmospheric nitrogen. Fixation of nitrogen by combustion could be possible in engines under high temperature. Nitrogen (N) in the air may be transferred to oxides of nitrogen by lightning. In developing countries, like Bangladesh, oxides of nitrogen, mainly nitric oxides (NO) come from the transfer of N during burning of vegetation. There is also a large contribution from microbial oxidation of ammonium in soils. The combination of fire-derived NO and hydrocarbons with sources of natural origin can result in significant production in ozone (O_3) as well. McElroy and Wofsy (1988) stressed that these conditions are increasingly common in the tropics.

Nitrous oxide (N_2O) is removed from the atmosphere mainly by photolysis. The rate of removal was calculated by McElroy and Wofsy (1988) as $10.5 (\pm 3) \times 10^6$ tons N year^{-1}. The rate of increase in the atmosphere is 0.7 ± 0.1 ppb/year (3.5×106 tons N year^{-1}). Therefore, total emissions for the amount of N_2O increased to $14 \pm 3 \times 10^6$ tons N year^{-1} as shown in Table 3.6.

Table 3.6 Nitrogen oxide concentration in the atmosphere

(A) Atmospheric burden (10^6 tons as N)	1,500
(B) Sinks + accumulation (10^6 tons as N/year)	
Stratospheric photolysis + reaction with [O]	10.5
Accumulation (0.7 ppb/year)	3.5
Total	14.0
(C) Sources (10^6 tons as N/year)	
Ocean	2.0
Combustion: coal + oil = 4 ± 1; biomass 0.7 ± 0.2	4.7
Fertilized agricultural lands	
Grass lands	8.1
Boreal and temperate forests	0.1
Tropical and subtropical forests and woodlands	7.4
Total	15.6
(D) Tropical contribution (ppb)	
Burning	100
Forest clearing	160
Oxidation of hydrocarbons	150
Total	410

Source: McElroy and Wofsy (1988)

Nitrous oxide is a free obligatory intermediate in denitrification, for example,

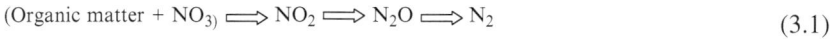

$$(\text{Organic matter} + NO_{3)} \implies NO_2 \implies N_2O \implies N_2 \tag{3.1}$$

Such a sequential reduction of nitrogen atom provides a respiratory path for a wide variety of bacteria under anaerobic conditions making the nitrogen available to the plants. Hence, availability of natural nitrogen partly depends on the N_2O (last step of the chain) concentration in the atmosphere. It is worth mentioning here that nitrogen availability may alleviate the aerial fertilization through carbon. In a subsequent chapter, we will discuss the issue. EPA (2010) reported that global N emission from natural sources ranges to 6.6 Tg in the form of N_2O.

Soils in tropical forests emit N_2O at rates far in excess of what is seen in most other environments. For example, release rates for tropical forests average about 2×10^{10} molecules cm^{-1} s^{-1}, as compared to $1–2 \times 10^9$ in temperate soils (McElroy and Wofsy 1988). Denitrification is most often observed in environments isolated from atmospheric oxygen and supplied with adequate oxidizable detrital materials, organic rich sediments, flooded soils, and closed ocean basins.

An interesting inverse relationship between fluxes of CH_4 and emissions of N_2O was also observed. Sites that consumed atmospheric CH_4 had the highest emission rates for N_2O; sites that emitted CH_4 had lower emissions of N_2O. The latter tended to be waterlogged or nearly so. However, microbial organisms play a critical role in the emission of nitrogen oxides. Microbial production is usually an anaerobic process, whereas consumption of CH_4 requires oxygen. The large emission of N_2O may be interpreted as an indication of rapid oxidation of mineral nitrogen in tropical forest soils, which emphasizes the idea that the nitrogen cycle in tropical forest is less conservative. This means that aerial fertilization effects due to CO_2 increase in the atmosphere will not be as effective in tropical forests as in temperate forests. This also means quite a lot in the C-trading system, specifically that the expected growth from tropical areas will not be achieved unless adequate N fertilization is made. However, we have to remember that if we apply N, we are changing soil environment significantly.

3.2.4 Anthropocentric Source of N

Where microbial emission of nitrogen oxide is a natural process, anthropocentric emission is a burden to nature and, hence, contributes more to the greenhouse effect. Use of nitrogenous fertilizer would artificially enhance the bioorganic emissions of N_2O leading to increasing concentration in the atmosphere. McElroy and Wofsy (1988) mentioned that about 0.1–0.5 % of the reduced N in fertilizer is converted to N_2O within a few weeks of application. The ultimate release of N_2O could be higher, however, since fixed nitrogen is likely to be assimilated into organic material and reoxidized a number of times before it is lost from the soils. This emission is in addition to the combustion source of anthropocentric N_2O emission. The reaction that is responsible for emission is

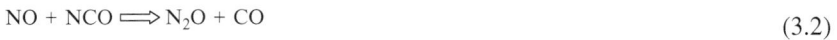

$$NO + NCO \Longrightarrow N_2O + CO \tag{3.2}$$

McElroy and Wofsy (1988) cited that approximately 10 % of the nitrogen in fuel is converted to N_2O during a typical combustion process. The oceanic source of N is usually derived from the nitrification process, using observations of accumulation N_2O and depletion of O_2 in marine waters supported by extensive data for N_2O dissolved in surface waters of the world's oceans. The value for combustion is usually derived from data defining the composition and utilization of various fuels, assuming that 10 % of fuel N is converted to N_2O. The N in fertilized agricultural lands can be based on the assumption of chemical fertilizer application at the rate of 40×10^2 tons N per year and a similar rate for the use of other manures with an overall yield of $1 + 0.5$ % N_2O. The contribution of forest toward N can be calculated as an average value of the direct measurement of the soil. From the discussion above, it reveals that the natural emission of N remains the interplay of green biomass with nitrogen richness of soil. However, from the table, we see that anthropocentric origin of N emission is significant.

Kopacek and Posch (2011) estimated that N emission from anthropogenic sources is about 30 % of total emission. From Table 3.6, it also appears that anthropocentric emissions account for about one-third of the total N emission. If the present pattern of emission continues, the abundance of atmosphere N_2O should slowly increase (the level in 1984 was 303 ppb). There is little reason to project that the emission will remain constant in the future. Sources associated with combustion and intensive agriculture are likely to increase and disturbed exploitation of tropical forest may also increase in future contributing increased fluxes of N_2O in the atmosphere. A much improved study and understanding is needed to predict future trends of the emission of N_2O. Phoenix et al. (2006) have shown that due to excessive N deposition in tropical and subtropical areas, biodiversity in 34 hotspots of the world would be at risk of degradation.

3.3 Greenhouse Gases and the Ozone Layer

The terrestrial ecosystem and atmospheric chemistry could have a cross relation, the understanding of which is essential for obtaining clues to the nature of the accumulation of greenhouse gas in the atmosphere and the impact of this on vegetation growth. The atmospheric O_3 layer is biologically beneficial to plants because it protects the green biota from the harmful effects of the sun's ultraviolet rays in the range of 310–240 micro-m length. Most greenhouse gases are active in reducing the O_3 layer in the atmosphere through a series of reactions.

3.3.1 Reaction of Greenhouse Gases with O_3

CFCs were assumed to be the most reactive and harmful to O_3 because they are very light gas and quickly escape to the stratosphere, and the reaction of chlorine (Cl) with O_3 oscillates for an indefinite period of time. CFC emissions are no longer critical because emission of CFCs has now stopped across the world. However, other greenhouse gases also react with O_3 under suitable situations. N_2O (nitrous oxide) is formed by microbial activity in soils and aquatic systems, a by-product of both oxidation (nitrification) and reduction (denitrification) of fixed nitrogen. McElroy and Wofsy (1988) mentioned that the total nitrogen (N) emission from the global source is about 10^7 tons/year with limit of $+30\%$. N_2O reacts in the following way:

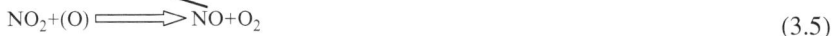

$$N_2O + (O) \Longrightarrow NO + NO \tag{3.3}$$

$$NO + O_3 \Longrightarrow NO_2 + O_2 \tag{3.4}$$

$$NO_2 + (O) \Longrightarrow NO + O_2 \tag{3.5}$$

The arrowheads in the equation demonstrate that in the presence of O_3, the reaction can oscillate a long time if N_2O is not dissociated by sunlight breaking the O_3 into O_2. Methane (CH_4) also has an important role in the chemistry of the stratosphere over and above its role as a sink of chlorine. McElroy and Wofsy (1988) described this as follows:

$$Cl + CH_4 \Longrightarrow HCl + CH_3 \tag{3.6}$$

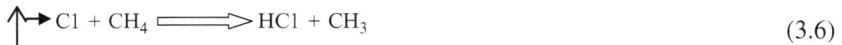

but HCl is unstable in presence of H_2O group and the reaction is:

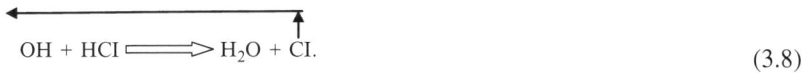

$$H_2O + [0] \Longrightarrow OH + OH \tag{3.7}$$

$$OH + HCl \Longrightarrow H_2O + Cl. \tag{3.8}$$

Here, we wished to show with an arrowhead how, in the presence of ozone and moisture, Cl can attack methane again and again. However, the abundance of water in the stratosphere is very little especially in tropical areas where the tropopause is not cold enough to hold water and convect it to the stratosphere. So, though most CH_4 is emitted from the marshy areas and rice fields of tropics, it may often be beneficial for the environment. Similarly, if CO_2 is present in the stratosphere, it may tend to cool the stratosphere, altering rates for several key reactions, offsetting partially the reduction of the O_3 layer and hence giving more protection from UV rays to green plants (McElroy and Wofsy 1988). But as CO_2 is a heavy gas, it may be very scarce in the stratosphere.

Oxidation of CO and CH_4 and other hydrocarbons can provide an important source of H bearing radicals in the atmosphere, leading either to production or destruction of O_3. Hydrogen-bearing radicals react with NO producing NO_2 which dissociates easily by light into NO and [O] atoms. The [O] atoms react with O_2 and form O_3. The reaction can be shown as follows:

$$[O] + H_2O \Longrightarrow OH + OH \tag{3.9}$$

$$OH + CO \Longrightarrow H + CO_2 \tag{3.10}$$

$$H + O_2 + M \Longrightarrow HO_2 + M \tag{3.11}$$

$$HO_2 + O_3 \Longrightarrow OH + O_2 + O_2 \tag{3.12}$$

$$or\ HO_2 + NO \Longrightarrow NO_2 + OH \tag{3.13}$$

$$NO_2 + heat \Longrightarrow NO + [O] \tag{3.14}$$

$$[O] + O_2 \Longrightarrow O_3 \tag{3.15}$$

The above reactions show the role of presence of moisture/precipitation in the atmosphere. It also shows the way that the production of O_3 in the atmosphere depends on the availability of NO. If the abundance of NO is high, for example, McElroy and Wofsy (1988) mentioned if it is above 70 ppt (parts per trillion), as it is above cities and over large regions of continents, oxidation of CO and hydrocarbons leads to net production of O_3; otherwise, removal of CO and hydrocarbons through the natural process of oxidation will result in consumption of O_3. But we have seen in the reaction (3.3, 3.4, and 3.5) that NO is a destructor of O_3. Probably the presence of CH_4 may reverse the process, which we could express in a simple way as Lovelock's (1988) Gaia statement "nature is self-corrective." However, the dangers lie that this O_3 formation could be ambient and likely to be carcinogenic to human health.

3.4   Perception of Greenhouse Effect

During the 1970s and 1980s, a considerable development took place in the understanding of global atmospheric chemistry. It was enunciated that besides CO_2, several other gases are radiatively active in the atmosphere, the so-called greenhouse gases, and these were steadily increasing in concentration (Table 3.1). Estimates of their future level suggest that, in the next few decades, they might together contribute to any greenhouse warming as much as CO_2 (Pearman 1990). The net effect of this could be a reduction in the time required for significant warming. The time required for change in the climatic patterns of the world might be reduced to half what would have been expected with the increase of CO_2 alone.

Despite this, there is a school who believes that there is no substantial evidence that the earth is warming up. Using the NASA data for 10 years (1979–1987), Spancer and Christy (1990) have stated that satellite data studied by NASA have shown no evidence of greenhouse effect. The trend of global temperature change has not been continuous. Further analysis of satellite data has shown an adequate coverage of global temperature change and has similarity with the global meteorological data. The authors believe that the media has failed to appreciate that atmospheric scientists developed the concept of "greenhouse effect" primarily on theoretical grounds and not on the basis of a short period of observation. Hence, the concept of greenhouse effect may not be rejected on the grounds of satellite data. However, the data from the satellite may not be reliable because sometimes data may be affected by the presence of clouds on the basis climate events happening now.

In the 1970s and 1980s, there has also been major progress regarding atmospheric chemistry and the role of trace gases. A list of trace element gases having radiative forcing effect is provided in Table 3.1. The detailed effect of trace elements on the greenhouse effect is still not clear. So, it is not possible on this basis to predict the nature and extent of greenhouse effect. Tuker (1990), however, suggested that the General Circulation Model (GCM) is a good point to start to analyze the regional impact of trace gases.

Fossil fuels are said to be the major source of greenhouse gases. The world generates 62 % of its electricity from fossil fuels, mainly in the form of coal. This represents 27 % of the total carbon emissions from fossil fuels (Reid 1990). All fossil fuels are not equal contributors to the greenhouse effect. For example, for the same amount of useful energy, oil emits 38–43 % and coal 72–95 % more CO_2 than that of natural gas (Allen et al. 1990). Along with CO_2, consideration needs to be given to other gases before making any definitive judgment of fossil fuel sources. However, the IEA prediction was that the world's energy-related CO_2 emissions would be expected to rise by about 50 %. The fastest rise is expected to come from developing countries due to a sharp increase in the use of energy.

References

Allen RW, Clively SR, Tilly JW (1990) Global greenhouse and energy situation outlook. In: Swaine DJ (ed) Greenhouse and energy. CSIRO, Canberra, pp 47–57

Cleary D (1991) The Brazilian rainforest: politics, finance, mining, and the environment. Special report No 2100. The Economist Intelligence Unit, London

Dixon BF (1990) Methane losses from the Australian Natural Gas Industry. In: Swaline DJ (ed) Greenhouse and energy. CSIRO, Canberra, pp 137–146

EPA (2010) Methane and Nitrous Oxide emission from natural sources. EPA, Washington, DC, 194 pp

Falk JA, Brownlow A (1989) The greenhouse challenge: what is to be done. Penguin Book, Ringwood

Kopacek J, Posch M (2011) Anthropogenic nitrogen emissions during the Holocene and their possible effects on remote ecosystems. Global Biogeochemical Cycles 25: GB 2017, 16 pp

Lovelock J (1988) The ages of Gaia: a biography of our living earth. Oxford University Press, Oxford

McElroy MB, Wofsy SC (1988) Tropical forests: interaction with the atmosphere. In: Prance GT (ed) Tropical rainforest and the world atmosphere. International Book Distributors, Dehradun, pp 33–60

Pearman GI (1990) Energy in the greenhouse effect: setting the scene. In: Swaine DJ (ed) Greenhouse and energy. CSIRO, Canberra, pp 1–10

Phoenix GK, Hicks WK, Cinderby S, Kuylenstierna JCI, Stock WD, Dentener FJ, Giller KE, Austin AT, Lefroy RDB, Gimeno BS, Ashmore MR, Ineson P (2006) Atmospheric nitrogen deposition in world biodiversity hotspots: the need for a greater global perspective in assessing N deposition impacts. Global Change Biol 12(3):470–476

Rastetter EB, Ryan MG, Shaver GR, Melillo JM, Nadelhoffer KJ, Hobbie JE, Aber JD (1991) A general biogeochemical model describing the response of carbon and nitrogen cycles in terrstrial ecosystems to changes in carbon-di-oxide, climate and nitrogen deposition. In: Kaufmann MR, Landsberg JJ (eds) Advancing towards closed models of forest ecosystems. A special volume of tree physiology, Heron Publications, Victoria 9: 101–126

Reid AF (1990) Research and technology opportunities for energy related reduction of greenhouse gas emissions in Australia. In: Swaine DJ (ed) Greenhouse and energy. CSIRO, Canberra, pp 31–46

Spancer RW, Christy JR (1990) Precise monitoring of global temperature trends from satellites. Science 247:1558–1562

Tuker GB (1990) Scientific uncertainties associated with the greenhouse problem. In: Swaine DJ (ed) Greenhouse and energy. CSIRO, Canberra, pp 11–23

Williams DJ (1990) Australian methane fluxes. In: Swaine DJ (ed) Greenhouses and energy. CSIRO, Canberra, pp 165–176

Chapter 4
Effects of Climate Change on Vegetation

The effect of climate change on vegetation is expected due to changes in the temperature, rainfall, and climate pattern as a result of which nutrient cycles, microbial activities, as well as physiological activities of plants will vary. Changes in precipitation may change moisture regimes. Soil erosion, salinity, acidity, and other physical and chemical factors may be affected as well. All these factors are closely related to vegetation growth. For example, with global temperature projected to rise to 3°C by the year 2030, this could cause the height of the sea level to rise by 0.8–1.8 m over the next century mainly due to thermal expansion of water and the melting of glaciers. A 1-m rise in sea level could jeopardize one-third of the world's cropland either through direct flooding or through salt water intrusion and would enhance the emission of greenhouse gases further.

CO$_2$ is considered as most responsible gas for climate change. Most CO$_2$ emission results from the burning of fossil fuels (Table 3.3). The emission from forest clearance is also significant. Deforestation in the Amazon alone contributes about 1 billion tons of CO$_2$ each year. Forest clearance also releases other greenhouse gases such as CH$_4$ and N$_2$O. Some scientists estimated that CH$_4$ is a ten times more powerful greenhouse gas than CO$_2$, and it is increasing at the rate of 1.2% per year in the atmosphere. There are other sources of CH$_4$; for example, the cattle population is growing by 1% and rice and paddy production is increasing by 0.7%. N$_2$O is also a very strong greenhouse gas and resides in the atmosphere for about 170 years, and its level is also increasing by 0.4% a year (Anon 1991). These are all greenhouse gases causing greenhouse effect and hence responsible for climate change. The consequences of the greenhouse effect on biomass growth could be through the following:

- The reduction of agriculture due to area losses under flood and salinity and due to dry out actions of increased temperature.
- The rate of extinction of species will be greatly accelerated as habitats will change faster than the ability of many species to adapt.

M. Ali, *Climate Change Impacts on Plant Biomass Growth*,
DOI 10.1007/978-94-007-5370-9_4,
© Springer Science+Business Media Dordrecht 2013

Table 4.1 Plant and ecosystem responses to CO_2 concentration (*Source*: Strain 1985)

1. First-order effects		2. Second-order effects	
(A) Primary effects	(B) Secondary effects	(A) Primary organism interactions	(B) Secondary organism interactions
1. Photosynthesis	1. Photosynthate concentration	1. Plant-plant interaction	1. Evolutionary effects – genetic differentiation
		(a) Interference	
	2. Photosynthate composition	(b) Competition	
2. Photorespiration		(c) Symbiosis	
	3. Photosynthate translocation	2. Plant-animal interactions	
		(a) Herbivory	
3. Dark CO_2 fixation	4. Growth rate	(b) Pollination	2. Ecosystem responses integration of all effects
		(c) Shelter	
	5. Growth form	3. Plant-microbial interaction	
4. Stomatal aperture	6. Reproduction	(a) Disease	
	7. Plant water status	(b) Decomposition	
	8. Tolerance to gaseous atmospheric pollutants	(c) Symbiosis	

- Hundreds of millions of people will be at direct risk from flooding, especially in the densely populated river deltas of Egypt and Bangladesh. They will be encroaching on new land and transform them to their habitats.

The changes may occur as direct or indirect consequences of greenhouse gases. By direct change, we mean the greenhouse gases will directly be involved in influencing the growth of biomass, and by indirect influence, we mean that the gases will change something else like temperature or microbial growth which will eventually influence the biomass growth. Both direct and indirect influences could happen at the same time. According to Strain (1985), following (Table 4.1) responses of ecosystems to the sequestering process of carbon dioxide enrichment may happen.

First-order effects are based mainly on individual responses to and individual behavior in the enriched CO_2, whereas the second-order effects are ecosystem responses to enriched CO_2. In both orders, some changes are "primary," that is relatively easier to detect, and some are "secondary" and very complex to explain. Such changes may induce a change in the relative composition of greenhouse gases in the atmosphere and, hence, may change the extent of greenhouse effects.

Among the greenhouse gases, CO_2 has a direct relation with vegetation growth through the process of photosynthesis. Hence, growth responses of plants will vary with the variation of CO_2 concentration in its environment. The responses of plants to increased CO_2 concentration can be described at a molecular as well as an organ level or whole plant level. Description of responses at a molecular level has not been discussed here because the process is limited to a tiny fraction of time and the field is still to be explored. On the other hand, description at the whole plant level may also be unsatisfactory because the growth of a whole plant is a complex combination of a multi-process model (photosynthesis, respiration, etc.) and organ model (partitioning of root, stem, and leaf). The whole plant model would be more appropriate to describe the responses of whole vegetation, that is, to describe a general vegetation purpose model. The response of the general vegetation varies with the climate and the vegetation type, such as native forest, grass land, desert land, etc., the responses of which to an increased CO_2 concentration are yet not clear. Hence, the responses have been described here as sub-models segregating into process model, organ model and soil model.

4.1   The Effect of Increasing CO₂ on Vegetation

Carbon is an essential element of life on earth. The biochemical energy necessary for continuation of life is carried in the atomic bonds of the carbon compounds of organisms. This carbon is then transferred from organisms to organisms. At each step of transfer, a proportion of carbon is usually returned to atmosphere as CO_2. In the very beginning, carbon is mainly available in green plants in the organic form of CO_2. This gas moves from the atmosphere, where it is very diffuse (3 in 10,000 or 300 ppm), into green plants where the carbon atoms are concentrated (4 in 10) in organic compounds (Strain 1985).

CO_2 is a natural component of our atmosphere and is partially responsible for keeping the earth at a temperature that supports life. CO_2 is also a nutrient for green plants and is a by-product of human metabolism. Concentration of CO_2 has increased in the atmosphere, and if the increase continues, it may produce major global changes by:

1. Increasing global temperature with subsequent changes in other climate variables
2. Changing the earth's vegetation by affecting photosynthesis and the water use efficiency of plants

These changes may, in turn, result in other changes in the biosphere which we are calling here indirect changes. The indirect changes in climate and vegetation may have global and regional effects such as changes in sea level, water availability, geographic pattern of agriculture, forests, and fisheries. All these changes may interact with each other, thereby influencing the extent of individual impact resulting from these changes. Impacts of changes in one factor will partially balance the impacts of changes in others and vice versa; that is, the full extent of the effects of one factor cannot be described by isolating it from others. For example, changes in water resources, temperature, and rainfall are each related to the biomass growth, whereas they themselves are interrelated. These interactions may have a feedback to modify the direct effects of CO_2 on vegetation.

Vegetation response due to increased atmospheric CO_2 may be modified by the effects of climate change on vegetation and vice versa; that is, the effects will be synergistic. Changes in vegetation may feed back to modify climate by changing the rate of evapotranspiration from plants, evaporation of moisture from soil, or by changing surface roughness or albedo. The amount of CO_2 in the atmosphere may also be affected by CO_2-induced changes in vegetation. Growing plants take up CO_2 (sinks) and dead decaying plants give off CO_2 (sources). These highly interdependent CO_2-induced effects reveal that the whole system must be examined to predict the extent of direct effects of increased CO_2 on biomass growth.

Increased atmospheric CO_2 is likely to benefit plants. This is so because of increased photosynthesis and decreased transpiration. Decreased transpiration is related to the effect CO_2 has in reducing the size of the stomata, resulting in decreased water loss during transpiration. This effect is expected to at least partially offset the effects of increased droughts that may occur in some regions via CO_2-induced climate change. These aspects of raising CO_2 concentration in the atmosphere may be reversed, canceled, reduced, or enhanced by two factors:

1. Differential plant species responses to CO_2
2. Climate changes associated with increased CO_2

Some classes of plants are expected to benefit more than others from the direct vegetative effects of increased CO_2. In some agricultural systems, economically undesirable plants could benefit more than desirable crops (White 1985). In forests, the long time scale over which individual trees compete with each other may lead to a wide variety of outcomes. Nutrient stress may further complicate the assessment of differential response. In forests where forest soils are not directly controlled by the application of fertilizer, the direct effect of CO_2 increases may be limited and vary from species to species by the inability to obtain necessary nutrients.

Although CO_2 increases are likely to be relatively constant from one region to the next, the associated climate change will be spatially heterogeneous and consist of the integrated effects (positive or negative) of a myriad of variables. The interaction of climate change with direct vegetative effects of CO_2 further expands the number of possible vegetative responses, including beneficial and detrimental outcomes. For example, in the case of aquatic plants, an increase in CO_2 may benefit the plant through an increase in photosynthesis (Blasing 1985), but the effect on transpiration seems to be less important than that on the terrestrial plant species. The influence of climate will be complicated by the geographic diversity of climate change and by the effects of climate change on the ocean currents, which influence the water temperature and nutrient supply. Thus, the details of different plant species responses and the effect of climate change on forest, agriculture, or aquatic systems are not very clear.

Response times are also uncertain. The long life span of trees implies that forests might take as long as a century or more to reveal a response to an appreciable climate change. Therefore, a GCM simulation may not be appropriate to predict the growth effect of CO_2 concentration. A succession model as suggested by Solmon and West (1985) could describe the growth effect better. Although the impact of increased CO_2 on vegetation has been described, it is important to note that the vegetation response may partially counteract rising levels of atmospheric CO_2 and consequent climate effects. Forests, for example, are an important component of the global carbon cycle, and the globally integrated effects of increased atmospheric CO_2 may lead to changes in the amount of carbon stored in the world's forests. If only direct vegetative effects are considered, the effect is likely to be an increase in carbon stored in forests, although this effect will probably be partially offset by increased respiration. The geographic distribution and seasonal timing of CO_2-induced climate changes will also influence the aspect of this carbon budget (Blasing 1985). Thus, even the direction of future changes in total carbon stored in the world's forests is not certain.

Carbon dioxide contributes through photosynthesis, to the production of carbohydrates of which about 50% is carbon (Maclaren 1991). Though CO_2 is so important for trees, it comprises only 0.035% of the atmosphere. Thus, it may not be surprising that a small increase in atmospheric CO_2 can trigger large growth responses. A doubling of current CO_2 levels is expected to increase plant productivity by one-third (Maclaren 1991). The processes involved in such an increase in plant productivity are described in the following paragraphs.

4.1.1   Aerial Fertilization

The growth effect of CO_2 on plants is known as "aerial fertilization," and there are prospects for application of this fertilizer at a high rate on a global scale. Many laboratory and field studies confirm this statement. Some commercial nurseries are using this knowledge in artificial enrichment of atmospheres in greenhouses producing crops and vegetables. This claim, however, may sometimes be countered by the theory of "limiting factors."

If, for example, water or light nutrient is limiting to growth then a CO_2 increase could have a negligible effect on growth. Alternatively, CO_2 enhancement may actually alleviate water stress and possibly other types of stress as well. A 330-ppm increase in atmospheric CO_2 concentration is expected to halve the amount of water required to produce one unit of dry matter (Brown 1991). Thus, water-stressed trees can be expected to respond to a high CO_2 level proportionately more than that of non-stressed trees. Some scientists argue that a similar phenomenon applies to a situation where nitrogen or phosphorus is limiting because of CO_2-induced stimulation of both fine roots and associated mycorrhiza.

Naturally, if a particular stress is extreme enough, no amount of CO_2 enrichment will relieve it. The effect of aerial fertilization due to past increase in atmospheric CO_2 concentration can be studied by examining the growth rings of trees in old growth forest. The evidence of CO_2 fertilization has already been discovered, for example, in a virgin stand of long leaf pine (*Pinus palustris*) in Georgia, USA. A similar result has been attained with Bristlecone pine (*P. aristata*) (LaMarche et al. 1984), with the annual ring being 73–106% wider than expected.

The authors also mentioned that in Scots pine, the effect of enhanced CO_2 was reflected in a basal area increment 15–43% greater than expected.

The most direct evidence for global CO_2 fertilization comes from seasonal changes in the Northern Hemisphere CO_2 level. In each spring and early summer, atmospheric CO_2 concentrations drop as new growth takes up carbon, and there is a corresponding rise in CO_2 levels in late summer and autumn as deciduous vegetation sheds leaves and annual plants die and release their carbon.

Brown (1991) reported that global CO_2 concentration is increasing by about 1.5 $\mu l/l$ (microliter per liter) annually. Plant productivity is often stimulated by short-term exposure to a CO_2-enriched condition. Thus, the productivity and net uptake of CO_2 are likely to increase, but unmanaged forest vegetation (weeds!) might increase as the global atmospheric CO_2 concentration increases, provided that other factors remain constant and short-term effects of CO_2 enrichment may be maintained over longer periods.

Richmond et al. (1982) grew *Chlorella vulgaris* and *Spirulina platensis* together in the presence of varying concentrations of dissolved CO_2 and bicarbonate (HCO_3). He observed that *C. vulgaris* did better under high levels of CO_2 and low (HCO_3) conditions; whereas *S. platensis* dominated when the CO_2 concentration was low and the (HCO_3) concentration was high. Although the concentration of dissolved CO_2 in water depends on many factors other than the concentration of CO_2 in the atmosphere, this result nevertheless indicates the possibility that even aquatic systems may be directly affected by increasing atmospheric CO_2 concentration.

4.1.2 Physiological Processes Associated with Enriched CO_2

Trees are long-lived and large when mature. It is not clear to what extent the long-term growth responses of trees to CO_2 enrichment can be inferred from short-term studies of seedlings. Studies on the influence of atmospheric CO_2 on growth have been conducted by exposing seedlings to increased CO_2 for a short period. A small increase in relative growth rate (RGR) in response to CO_2 enrichment would result in a large increase in biomass, when compounded over several years. However, the enhancement of RGR in response to CO_2 enrichment may not persist when the duration of exposure is increased. Time-dependent decreases in the RGR of CO_2-enriched plants, relative to controls, have been related directly to a decrease in the net assimilation rate (NAR) (Tolley and Strain 1984) and to the onset of self-shading (Poorter et al. 1988). Alternatively, CO_2 enrichment could stimulate a requirement for mineral nutrients in excess of that supplied. A decrease in nutrient status could then lead to a decrease in RGR by causing a decrease in either NAR or the allocation of biomass to leaves. Thus, a decrease in RGR of CO_2-enriched plants relative to controls may be a direct physiological response to CO_2 enrichment or an indirect response dependent either on accelerated development or maturation or the accentuation of other growth-limiting factors. Prediction of tree growth responses to CO_2 enrichment requires an understanding of the relative importance of these different factors. Brown (1991) reported that the augmentation of growth under the CO_2-enriched situation was not held for more than 10 days at a specific level of nutrient supply; the color of leaves subsequently changed indicating cellular deficiencies of N, P, K, and salt associated with the observed growth responses to atmospheric CO_2 concentration.

4.1.3   Effect of Enriched CO_2 on the Photosynthetic Process

Diurnal and seasonal variations in irradiance and in soil and atmospheric variation cause continual changes in the physiological activity of plants. Plant biomass growth depends on the resultant of photosynthesis and respiration. The fixation of CO_2 depends on the photosynthesis of green plants. Photosynthesis itself depends on many other factors, which may need to be considered to identify the effects of increased CO_2 on the photosynthetic process. For example, phosphate is used in the reaction of CO_2 fixation. A portion of this phosphate is supplied from chloroplasts. Sharkey and Vandeveer (1989) reported that phosphate within the chloroplast stoma is a substrate in ATP synthesis, and this regulates starch synthesis; and in the cytoplasm, inorganic phosphate is required for sucrose synthesis. Phosphate utilization in enzyme-mediated reactions of carbon fixation may, therefore, cause local phosphorous deficiency, inducing feedback inhibition of photosynthesis during some parts of the diurnal cycle. According to Woodrow and Berry (1988), the accumulation of photosynthesis in the leaf mainly affects starch-sucrose partitioning and has little effect on CO_2 fixation except under conditions of water stress or at a low temperature, when CO_2 fixation may decline as assimilates accumulate. This end product inhibition of photosynthesis may be exerted through an effect on the activation of ribulose 1,5 bisphosphate carboxylase oxygenase, although there are a number of other possible mechanisms (Luxmoore 1991). Sasek et al. (1985) reported that an increase in photosynthetic rate has been seen in association with the depletion of leaf starch. It seems that though the CO_2 concentration increases, photosynthesis may not increase the starch content in the leaf. But an increase in biomass production may then be adjusted by other processes, for example, increasing leaf area or leaf number. If this is true, then it is clear that there is a limit to the way the tree may act as sink for CO_2.

4.1.4   Individual Plant Response to Photosynthetic Change

The effect of CO_2-induced change in photosynthesis is positive to plant growth in general; however, the rate of growth may vary from plant to plant depending on the rate of photosynthesis, timing of photosynthesis, and the component to which the photosynthates are distributed.

Table 4.2 Percentage
increase in total biomass for
growth 28 days after planting
at different CO_2
concentration relative to the
values at 300 ppm

Species	Biomass increase in %	
	600 ppm	1,200 ppm
C_3 Plants		
Datura stramonium (1)	74	115
Datura stramonium (2)	60	107
Chenopodium album	76	140
Polygonum pensylvanicum	48	100
Abutilon theophrasti	38	65
Ambrosia artemisiifolia	68	112
Acer saccharinum	32	63
Populus deltoides	29	20
Platanus occidentalis	33	33
Glycine max	47	100
Helianthus annuus	40	55
C_4 Plants		
Setaria faberi	42	106
Setaria lutescens	70	45
Amaranthus retroflexus (1)	36	49
Amaranthus retroflexus (2)	29	48
Zea mays	21	10

Source: Bazzaz et al. (1985)

4.1.4.1   Effects due to Changes in the Rate of Photosynthesis

Different types of plants may have different rates of photosynthesis. For example, C_3 and C_4 plants have different response to photosynthesis and transpiration, and hence, enhanced growth on those types of plants is expected to vary from each other. Bazzaz et al. (1985) cited about 14 species studies which show a wide variety of changes in plant weight and the photosynthetic rate for two CO_2 concentrations. Table 4.2 shows the details of comparatives.

Although the choice of CO_2 concentration in the experiment is much higher than the expected range of CO_2 concentration in the atmosphere almost double and quadruple, the comparatives can be taken as indicator in variation of responses in different plants. In average although C_3 plants have shown much more productivity in higher concentrations, in some cases almost doubled, species like *Platanus occidentalis* did not show any change in the C_3 group. In practice, *Populus deltoides* has shown a decline in growth from 29 to 20%. On the other hand among the C_4 groups, rate of increase in *Setaria lutescens* and *Zea mays* has decreased significantly. Interestingly in the same genus *Setaria*, the rate of biomass growth has increased more than double for one species, whereas in another species it has been reduced to almost half. These variations indicate that besides CO_2, there are factors that are regulating plant growth productivity and such factors would bring potential uncertainty in the productivity of agricultural crops, which are mostly C_4 plants.

Table 4.3 Effects of CO_2 concentration on dry weights of corn, itchgrass, soybean, and velvetleaf harvested at 12, 24, and 45 days after planting

Species	Planting day	Dry wt (g) at different CO_2 concentration		
		300 ppm	600 ppm	1,000 ppm
Corn	12	0.63	0.73	0.76
	24	6.96	6.24	6.63
	45	91.29	89.49	80.08
Itchgrass	12	0.08	0.16	0.15
	24	2.10	3.82	3.50
	45	39.25	47.47	38.62
Soybean	12	0.34	0.52	0.61
	24	3.60	4.68	6.38
	45	50.55	62.19	87.09
Velvetleaf	12	0.08	0.272	0.21
	24	1.94	3.727	3.65
	45	35.34	47.96	54.34

Source: Bazzaz et al. (1985)

4.1.4.2   Effects due to Changes in the Timing of Photosynthesis

In the previous section, it has been shown that total biomass changes were significantly different among species. But the experiment based on 28 days growth may not be sufficient to comment on the impact. The impact of duration on the rate of growth is demonstrated in Table 4.3.

Table 4.3 shows the variation of responses of different species to different concentrations of CO_2 at different growth periods. For example, the growth of corn increased in the early 12-day growth period with the increase in the concentration of CO_2, but the growth gradually declined with the increase of CO_2 when the growth periods were extended for 24 and 45 days. The rate of declination increased with the increase in duration of the growth periods. In case of velvetleaf, though the growth was increased for 12- and 24-day growth periods for CO_2 concentrations of 300 and 600 ppm, the growth was decreased at 1,000 ppm. When the growth period was extended to 45 days, a substantial increase in growth was achieved even at 1,000-ppm concentration of CO_2. For itch grass, growth period through responses was positive at 300 and 600 ppm concentration of CO_2, whereas soybean shows a consistent increase in growth at every stage and at every experimental concentration of CO_2. This reveals that appropriate comment on the responses of vegetation to the increase of CO_2 concentration may not be possible from a single study or from study of a limited period.

The work of Tolley and Strain (1984) has shown that the early growth of sweet gum (*Liquidambar styraciflua*) and loblolly pine (*Pinus taeda*) during the early phase of their development, in response to an elevated level of CO_2, was strongly influenced by the light and water status of the environment. The response was

Table 4.4 Effects of a atmospheric CO_2 on growth of *Liquidambar styraciflua* and *Pinus taeda* at 32 weeks after planting in controlled greenhouses with a natural photoperiod and simulation of outside temperature

Species	CO_2 conc. (ppm)	No. of branches	Stem weight (g)	Leaf weight (g)	Root weight (g)	R:S ratio	Total weight (g)
Liquidambar	350	1.8	2.1	1.5	6.3	1.7	9.9
styraciflua	500	1.7	3.4	1.6	9.2	1.8	14.1
	650	4.0	3.7	2.0	8.6	1.5	14.3
Pinus taeda	350	2.0	0.4	1.4	2.1	1.2	3.9
	500	2.7	1.0	2.3	2.9	0.9	6.2
	650	2.3	0.8	2.1	3.2	1.1	6.1

Source: Sionit et al. (1985)

species dependent, with loblolly pine responding substantially less than the sweet gum at all light and water levels. Sionit et al. (1985) investigated the development of sweet gum and loblolly pine growing for an entire growing season, mimicking field temperatures throughout the season, under three CO_2 level. Results are shown in Table 4.4. The results confirm those presented from Bazzaz et al. (1985) in the previous paragraph (Table 4.2). At the end of the growing season, both species had responded to an elevated CO_2 level, but sweet gum grew taller and produced more branches and more stem weight than did loblolly pine. This difference again suggests that changes in the seedling growth pattern may alter the competitive relationship of the two species.

Oechel and Strain (1985) demonstrated from a study on the effect of CO_2 concentration of 330, 600, 900 ppm on flowering and fruiting with particular attention given to the timing of flowering that an increased level of CO_2 advanced the flowering time of several populations of *Phlox drummondii*. Increased CO_2 may enhance the flower production rate, longevity, and the total number of flowers produced. In another experiment, Bazzaz et al. (1985) have shown the effect of increasing CO_2 on the reproductive biomass of *Datura stramonium* and *Abutilon theophrasti*. As CO_2 increased, the seed number did not increase much in *D. stramonium* and actually decreased in *A. theophrasti*. However, this decrease was offset by an increase in individual seed weight. Fruiting increased in *D. stramonium*, and *A. theophrasti* grew more quickly though the final biomass was little affected by CO_2 concentration.

These findings show that it is not only the vegetative growth, the flowering, and fruiting of the plants might get changed significantly. The variation in flowering and fruiting might get further affected by temperature, precipitation, and seasonal changes due to climate change impact. If such abrupt changes come in sexual organs of the plants, whatever may be the vegetative growth, the fertility of plant will be affected by differential maturity of male and female organelles of flowers and due to lack of fertilizing agents. Under the circumstances, the species is likely to become extinct.

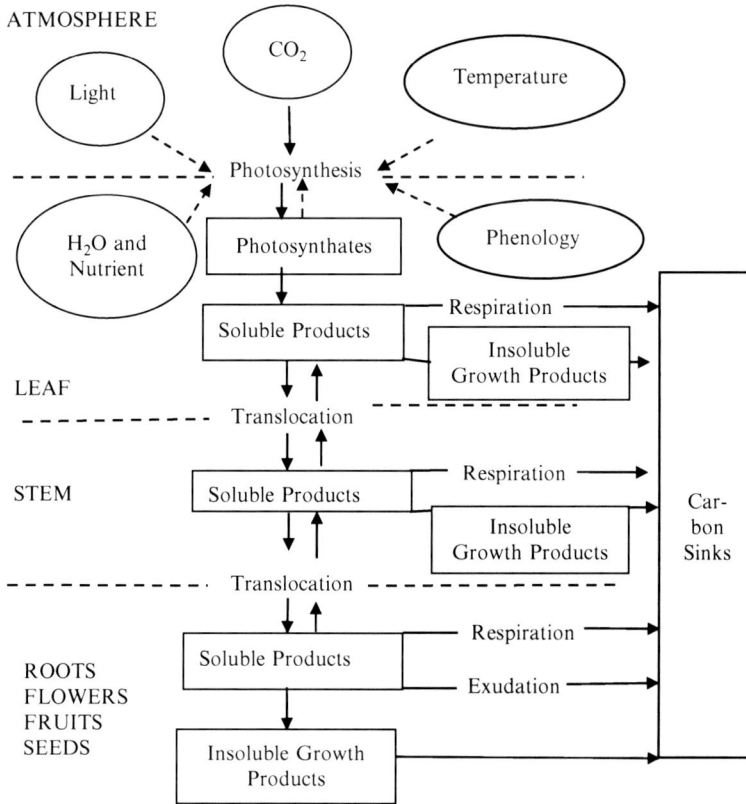

Fig. 4.1 Diagram of carbon (C) flow and utilization in plants. Specific interest is the question of inhibitory feedback of photosynthate production on C assimilation and questions of C distribution and allocation within plants as affected by atmospheric CO_2 enhancement. Also of interest is the relationship between atmospheric CO_2 concentration and C losses from plants. The C eventually leaves the plant in respiration, seed and fruit dissemination, root exudations, herbivory, senescence, organ abscission, or death (Source: Oechel and Strain 1985)

4.1.4.3   Effects due to Changes in Partitioning

While more carbon is fixed due to an increased level of photosynthesis, the site for the deposition of the carbon is a function of genetics of the plant involved, the environmental conditions prevailing at the time, and the phonological condition of the plant. Carbon fixed in photosynthesis may pass through a series of biochemical reactions that produce various metabolic or growth compounds. Figure 4.1 shows the probable sites for the carbon allocation.

Increased carbon, if allocated to the reproductive structure, may increase the reproductive potential of plants in a natural ecosystem. Such a response was observed by Oechel and Strain (1985) in *Chrysothamnus viscidiflorus*, a shrub, and also in some crops. Where carbohydrates are limited, this pool of nonstructural carbohydrates could result in increased growth. When a plant has excessive carbohydrates, little increase in growth may be expected due to increased CO_2 concentration in the atmosphere. Let us describe how the growth of organs takes place due to differential partitioning of accumulated carbon.

4.1.5 Allocation of Carbon to Shoot

Carbohydrates in plants are produced in the shoots and then distributed to the different parts of the growing region. The proportion of the carbon translocated to the different parts depends on the environmental conditions under which the parts are growing and the importance of that component to the plant. The distribution response or carbon in different components of shoots can be described as follows:

4.1.5.1 Leaves

The major portion of carbon fixation takes place on the surface of the leaf, and, hence, carbon compounds for tissue formation and growth are readily available in leaves after they become photosynthetic. Initial cell division and growth of leaves from leaf primodia in buds of vegetative plants require the translocation and essential mineral nutrients from elsewhere in the plant. Hence, leaf growth depends on the store of carbohydrates of the plant. It should be remembered from our earlier statements that enhanced leaf growth at elevated CO_2 varies from species to species and even from plant to plant depending on the environmental condition. Enhanced leaf development can be a problem if the evaporative leaf area increases more than the absorptive root area.

4.1.5.2 Leaf Area and Number of Branches

As leaf area per shoot and the number of branches increase in response to elevated CO_2 (Oechel and Strain 1985), the total number of leaves per plants increases, and the LAI (Leaf Area Index) may also increase. If the LAI increases and, particularly, if photosynthesis quantum is higher at high CO_2 concentration, net photosynthesis may increase within ecosystems. CO_2-induced changes in LAI may affect light penetration into the ecosystem, and, hence, LAI changes will change the distribution of light at different canopy levels. Consequently, the change in total photosynthesis in the whole system becomes uncertain resulting in the changes in undergrowth diversity composition of the ecosystem. If the LAI increases, it is likely that the branching of responsive trees will also increase to optimize the trapping of light.

4.1.5.3 Stem Growth

At high CO_2 concentration, plants frequently show higher rates of height growth. Perennial plants and annual plants with indeterminate growth became taller at elevated CO$_2$ concentration. Determinate annuals have shown to increase in height at a faster rate, but the final height attained may not be affected (Oechel and Strain 1985). Such a fast growth rate is important to avoid weed competition for temperate plants where the growth rate is relatively short. As the branching and their pattern may increase under elevated CO$_2$ situation, in some plants, increase in node formation under elevated CO$_2$ concentration may also be changed resulting in an unexpected change of wood quality.

4.1.5.4   Flower and Seed Production

An increase in the number of branches and the formation of an increased number of nodes may result in an increase in the number of floral buds if appropriate hormonal and environmental conditions prevail. We wish to remember here, as we said before, that there were experimental evidences that the seed size has been increased in certain crops due to increase in the CO$_2$ concentration. Though larger seeds imbibe water slowly, thereby, may show delayed germination, larger seeds may produce more vigorous seedlings resulting in faster development producing significantly more leaf area in the first few days of development. On the other hand, seed dispersal distance may be reduced due to increase in the seed size. As a result, community composition may be changed.

4.1.5.5   Asexual Reproduction and Tillering

One of the most important effects of an increase in CO$_2$ may be an increase in tillering (branching). This may occur even where there is a relatively small increase in photosynthesis. Such an increase in the number of heads of wheat was reported by Oechel and Strain (1985). Increase in branching may increase the asexual reproduction particularly for stolon-type plants. This might accompany by the changes of coppicing ability of some plants as well.

4.1.6   Carbon Allocation to Roots

The storage organ provides active sinks for the translocation of carbohydrates from the sites of formation in the leaves, and in that case, the end product inhibition may be minimal. With increasing atmospheric CO$_2$, most plants allocate proportionately more of the extra carbon below ground causing an increase in root shoot ratio. Differential allocation of carbon to various organs, while atmospheric CO$_2$ increases, will cause changes in the competitive behavior among plants. This competitiveness could be very important in the natural ecosystem. Plants that produce relatively larger roots will get an advantage in obtaining soil moisture and nutrients. On the other hand, changes in the root shoot ratio will change the balance of respiring to photosynthesizing tissue, and hence, the nature of the biomass growth may change.

4.1.6.1   Response of Root Growth to Enriched CO_2

Photosynthesis is essential for root growth. Van den Driessche (1991), in the experiment with 1-year-old Douglass fir seedlings, found that when the supply of freshly prepared photosynthesis was stopped, root growth was also significantly reduced. This means that where the CO_2 concentration is reduced, resulting in a reduction of photosynthate production, root biomass growth may also be low. Conversely we can say that in the greenhouse condition, when the CO_2 concentration is high, root production may increase, and, as a result, there may be a greater nutrient stress on the soil. On the other hand, a plantation may need to be more widely spaced to ensure healthy production under nutrient stress conditions. But it must be remembered that the rooting ability of different species will vary, depending on the age of the plant. Thus, it may not be possible to tell what will happen to different species which are already mature. In fact, root growth depends on many other soil physical and chemical factors which affect plant physiological processes. Plants with large carbon storage capacity in below-ground tissues (e.g., sweet potato, white potato) respond greatly with increase in below-ground production when CO_2 increases in the atmosphere (Bhattacharya et al. 1985).

4.1.6.2   Exclusion of Carbohydrates from Roots and Other Organs

In addition to increased growth associated with the increased carbohydrate supply, it has been proposed by Luxmoore (1981) that some of the extra carbon might be lost from the roots as soluble exudates. Root exudates are common and are known to have chemical and biological effect in the soil. Hence, increase in carbohydrates in the roots of leguminous plants may enhance symbiotic nitrogen fixation. Luxmoore (1981) hypothesized that any increase in root exudation could lead to increase in non-symbiotic nitrogen fixation, increased mycorrhizal activity, increased soil microbial growth and nutrient retention, and possible increased rates of mineralization. Additionally, if mycorrhizal hyphae are produced because of extra carbon supplies to the root, and if they extend out into the soil, thereby increasing absorption area, the water and nutrient supply to the host may be enhanced. The net effect might be an increase in plant growth via the mechanism diagramed in Fig. 4.1.

A wide variety of organic compounds are exuded from plant roots including carbohydrates, amino acids, organic acids, nucleotides, and enzymes (Rovira 1969). Among the organic acids, citric, fumaric, malic, oxalic, succinic, and acetic acids are important root exudates. In addition, there may be a considerable number of materials released by tissue sloughed by growing roots and by root decay (Oechel and Strain 1985). All these materials may be factors for causing allelopathic reaction and may initiate changes in community composition.

4.1.7   Other Responses

Excess carbohydrates might be distributed to components of trees other than the roots and shoots as follows:

4.1.7.1   Phenology and Senescence

Elevated CO_2 may have an effect on the phenology and senescence of plants. Annual plants may develop more quickly under elevated CO_2, reaching full leaf era, biomass, and flower and fruit production sooner than plants at ambient CO_2 (Paez et al. 1984). These plants may not be more productive than those normal CO_2, but maximum biomass is reached at an earlier time. In a competitive situation, early leaf seeds production could shift the population dynamics and competitive relations. However, Oechel and Strain (1985) referred to some studies which describe delayed senescence of some species. Delayed senescence potentially increases the growing season and increases production. However, it might also make plant vulnerable to frost damage in temperate areas and pathogenic or pest attack in tropical areas.

4.1.7.2   Tissue Quality and Chemical Composition

It may be expected that in a nutrient-limited environment, the carbon to nitrogen ratio in plant organs may increase with the increase of CO_2 concentration, while at the same time, other macro- and micronutrients may decline. Increase in structural carbohydrates and storage carbohydrates might be expected. Such trends could reduce decomposition, and, hence, a change may occur in the nutrient cycle and may affect further growth in the natural ecosystem. As a result, it is likely that the flux and speed of biogeochemical cycle could get changed.

4.2   The Effect of Increasing CO_2 on the Distribution of Organic Matter and Nitrogen Cycling

Consideration of the effects of enriched CO_2 on essential element in plant tissues is important because these elements are related to growth process. Growth processes need to be considered in determining the silvicultural treatment for a particular forest of a single species. Traditional silvicultural practices are based on production models which are derived in turn from field measurements of trees and statistical interpretation of their size, distribution, and changes in size over time. Such mensurational models are completely empirical and site specific. They describe tree growth, often simply as merchantable volume, on a particular site subject to particular soil and weather conditions. They cannot be used to examine the consequences of any significant departure from the conditions pertaining over the period of measurement, and they cannot simulate the consequences of changes in such factors as atmospheric CO_2 concentration, temperature, and water regimes. Because mensurational models are not generally driven by environment variables, they cannot be used to examine the influence of a wide variety of ecological and physiological process affecting the health and behavior of forest ecosystems. New variables should be added here, probably environmental variables, which may affect the normal physiological process contributing to changes in growth and population patterns. Landsberg et al. (1991) suggested that the CO_2 concentration in the atmosphere is one of the variables. So, on a wider scale, an appreciation of changes in atmospheric CO_2 concentration and their probable effects on global climates and factors such as atmospheric pollution make it essential to develop models capable of predicting the probable effects of these phenomena by measuring macroelements concentration in plant tissues.

Natural forests contain trees of many species and individuals of different ages. Those with closed canopies cover about one-fourth of the globe and constitute about 90% of the growing stock (Solmon and West 1985). These forests are generally densely populated, and the plants compete with each other for sunlight, water, and nutrients. Also, plants use different strategies to obtain the resources to grow and to reproduce in the presence of other plants. As a result, few trees in unmanaged forests may ever grow at their maximum potential rate or reach their maximum potential size. In the managed forest, the success of silvicultural treatments depends on the way they respond to these factors of environment. The environmental factors are expected to be influenced by CO_2-induced environmental changes as a result of which, under climate situation, stresses on tree growth may change significantly over time.

Other differences in response may arise because of the differences between the temperate and tropical ecosystems and the amount of woody tissue contained by the individuals. This may be greatly influenced by temperature. Biomass growth is the resultant of respiration and photosynthesis, and both will increase with the increase of temperature. Thus, the ratio of photosynthetic tissue to non-photosynthetic tissue can greatly influence the response of biomass growth. In the hardwood forest where there is large stock of woody tissue relative to foliage, increases in respiration are likely to be the dominant response to increase temperature, leading to a pronounced decline in net primary production (NPP). Rastetter et al. (1991) simulated a model on time course of NPP and soil respiration in Arctic tundra and temperate hardwood forest under exposure to a doubling CO_2 concentration, a 5°C increase in temperature, a combined CO_2 and temperature increase, and an increase in N deposition of 0.1 Nm^{-2} $month^{-1}$ over the growing season. The result revealed that NPP has marginally increased in tropical condition under combined situation only. This reduction in NPP may almost exactly cancel the initial fertilizing effects of CO_2 on broad-leaved forest when both temperature and CO_2 are increased. In the tundra, on the other hand, where there is very little wood relative to foliage, the initial response to increased temperature may be a small increase in NPP. This increase may amplify the initial fertilization effect when both temperature and CO_2 were increased.

Table 4.5 also shows that the tundra vegetation cycles N more tightly than does the hardwood forests. Where CO_2 and temperature is increased, but not N, plant uptake of N in the tundra was closely correlated with net N mineralization. Hence, as

Table 4.5 Changes in carbon (C) and nitrogen (N) mass and turnover rates in Arctic tundra (tund) and temperate hardwood (HW) ecosystems after exposure of 50 years to a doubling of CO_2 concentration, a 5°C increase in temperature, a combined CO_2 and temperature increase and an increased N deposition of 0.1 Nm^{-1} over the growing season

	Control		$[CO_2] \times 2$		Ambient (Amb) temp. +5°C		$[CO_2] \times 2$ + Amb temp. +5°C		Amb. N depo-sition + 0.1 N	
	C	N	C	N	C	N	C	N	C	N
Rates of C and N turnover (% year^{-1})										
Tund plant	23.7	19.7	22.7	20.2	23.8	19.6	22.8	20.2	24.5	20.3
Soils	0.6	0.2	0.6	0.2	0.8	0.3	0.9	0.3	0.6	0.2
HW. plant	6.4	18.6	6.1	15.6	6.3	21.1	6.8	20.7	6.8	19.1
Soils	7.1	1.7	7.1	1.6	8.7	2.4	9.2	2.4	7.3	1.8
Changes in C and N mass (% of control)										
Tund plant			113.0	91.3	151.1	158.6	172.6	146.5	118.4	116.7
Soils			101.2	100.1	100.5	99.5	102.1	99.6	102.0	101.0
Total			101.5	100.0	101.6	100.0	103.6	100.1	102.4	101.1
HW. plant			104.2	111.2	122.2	117.3	129.2	126.9	101.3	108.9
Soils			99.3	99.2	94.1	95.3	100.0	99.3	103.6	103.1
Total			101.9	100.2	108.8	97.5	115.3	101.6	102.4	103.6
C to N mass ratio (gC/gN)										
Tund plant	41.6		51.5		39.6		49.0		42.2	
Soils	17.5		17.7		17.6		17.9		17.6	
Total	17.7		18.0		18.0		18.3		17.9	
HW. plant	159.5		149.4		166.1		162.4		148.4	
Soils	13.2		13.2		13.0		13.3		13.3	
Total	25.4		25.9		28.5		28.9		25.1	

Source: Rastetter et al. (1991)

a result of these manipulations, there may virtually be no change in the total amount of N in the tundra vegetation. However, there may be some important shifts in the N distribution between vegetation and soil. Higher CO_2 concentration results in higher C/N ratios in plant materials and consequently could increase the N mobilization potential. Thus, the net effect of CO_2 in the tundra is to move N down into the soil organic matter. CO_2 has the opposite effect in the hardwood forest because the vegetation is usually better able to exploit efficiently a relatively fertile soil and because of the longer retention time of N within the vegetation.

4.3   Interaction of CO_2 with Other Elements

An increase in CO_2 concentration is not an isolated phenomenon but interacts with other elements depending on the environmental factors. This is shown by Brown (1991) who reported from his experiment on *Populus tremuloides* seedlings with three levels of N and two levels of CO_2 concentration (ambient 350 μl/l and enriched 750 μl/l) for 100 days. At a constant level of N, growth responses were initially quicker in enriched CO_2 than the ambient level, but ultimately the responses were similar at the end of the experiment. However, total CO_2-induced growth responses were greater at a higher level of N. It seems that at a higher CO_2 concentration, the use of N increases, and, as a result, the N pool becomes limiting and the growth response declines.

From here, one may assume that if fertilizer level is high, use of CO_2 will also be high. In fact, it is probable that any increase in the use of CO_2 will vary from species to species depending on the growth attributes of the trees and/or their genetic quality. It is impossible to evaluate the economic justification for fertilization of the forests throughout the world in order to achieve a major sink for greenhouse gas CO_2. Nevertheless, it is easier that N-fixing trees may act as greater sinks for CO_2 if their genetic qualities are improved.

Carbon dioxide enrichment and increasing availability of N increased the initial relative growth rate (RGR) as shown in the study of Rastetter et al. (1991) but shortened the period during which high growth rates were maintained (Table 4.5). It was found that at the highest level of N and enriched CO_2, the period of high RGR was shortest. At the lowest level of N and ambient level of CO_2 the period of high RGR was greatest. However, the peak of RGR was much higher at high N (0.19 g/g/day) than at low N (0.06 g/g/day). After a certain time, the relative growth rate declined below that of the lowest N level. It seems that the total N level was exhausted very rapidly when CO_2 concentration was high.

In order to estimate the relationships involved in this, Brown (1991) measured net assimilation rate (NAR), leaf area partitioning (LAP), and root mass partitioning (RMP). It was found that at both high and low N availabilities, an increase in CO_2 concentration increased RGR by increasing NAR. In the medium range of N, the RGR increased with both NAR and LAP coefficient. On the other hand, the relationships among RGR, NAR, and LAP were found to vary with nature of treatment and time. In high N seedlings, RGR decreased initially with the LAP;

Table 4.6 Changes in foliar concentration of N, P, K, Ca, and Mg in *Populus tremuloides* with time (mean ± one 95% confidence interval; $n = 5$; if no confidence interval is shown $n \leq 2$)

Treatment	Day of harvest	Concentration (mg/g)				
		N	P	K	Ca	Mg
15.5 mM-N						
350	40	42.0 (6.1)	4.7 (1.9)	16.6 (5.0)	1.8 (1.0)	2.4 (1.9)
	60	34.1 (5.8)	4.0 (1.6)	15.8 (3.9)	1.7 (0.6)	2.2 (0.7)
750	30	45.5 (8.9)	5.1 (1.4)	16.8 (7.3)	2.2 (0.8)	2.6 (1.2)
	40	27.4 (4.1)	3.0 (1.0)	13.2 (5.4)	1.0 (0.2)	1.4 (0.3)
	60	26.9 (6.2)	3.3 (0.8)	15.3 (4.7)	0.9 (0.2)	1.5 (0.3)
1.55 mM-N						
350	40	30.2	4.6	23.8	7.7	4.8
	60	18.9 (1.3)	2.2 (0.6)	20.4 (2.2)	2.1 (0.8)	3.0 (0.3)
750	40	25.0	4.3 (1.2)	19.3 (7.3)	5.1 (2.7)	3.7 (1.7)
	60	13.5 (1.3)	2.2 (0.6)	20.4 (2.2)	2.1 (0.8)	2.2 (0.8)
0.155 mM-N						
350	50	22.6		18.8	15.9	5.9
	60	22.7	3.7	22.6	15.1	6.0
	100	11.4 (2.6)	2.2 (0.6)	18.8 (5.9)	7.6 (5.1)	4.0 (1.6)
750	50	12.6	4.6	15.4	10.7	4.3
	60	12.6 (2.2)	2.7 (0.8)	20.0 (5.7)	7.2	4.4
	100	9.5 (2.2)	2.2 (1.2)	20.4 (1.4)	11.9 (8.00)	4.90020 (2.4)

Source: Brown (1991)
Plants were grown in atmospheric CO₂ concentration of 350 and 750 μl/l and fertilized with solutions containing 15.5 mM-N (macro nutrient concentration), 1.55 mM-N, or 0.155 mM-N

subsequently, both LAP and NAR decreased. In low $(N + CO_2)$ seedlings, the decrease in RGR was associated mainly with the decrease in NAR.

Brown (1991) also measured the concentration of N, P, Ca, Mg, and K in leaf as well as in the whole plant at different times and tries to find out their correlation with RGR and LAP performance. He found that in high $(N + CO_2)$ seedlings, the decrease in RGR and LAP was associated with decrease of N, Ca, P, and Mg both in the leaf and whole plant level. In contrast, in high N and ambient CO₂-treated seedlings, the decrease was associated with the decrease of N and P at the whole plant level only but not at the leaf level or the level of elements other than N and P. In low-N and high-CO₂ seedlings, the decrease in foliar P and whole plant N preceded the decrease in RGR. It was also identified that when foliar concentration of N decreased, the concentration of P, K, Ca, and Mg increased relative to N concentration, that is, an exhaustion of N content in the leaf may not cause a similar exhaustion of other minerals. Table 4.6 describes the situation.

The table shows the fall in foliar concentration of nutrients when duration of growth period increases which are important to visualize how harvesting period removes nutrient as foliar content. Notable though that in case of K, foliar content increases with increase in duration of harvesting. The table also shows that at higher concentration of CO₂, the foliar content of nutrient declines. This means that the green biomass can use nutrient more efficiently than that of lower C concentration.

References

Anon (1991) Rainforest destruction: causes effects and false solutions. World Rainforest Movement, Malaysia, Jutaprint

Bazzaz FA, Garbutt K, Williams WE (1985) Effect of increases of atmospheric carbon dioxide concentration on plant communities. In: Strain BR, Cure JD (eds) Direct effects of increasing carbon dioxide on vegetation. US Department of Energy, Washington, DC, pp 155–170

Bhattacharya NC, Biswas PK, Battacharya S, Sionit N, Strain BR (1985) Growth and yield of sweet potato (*Ipomea batatas*) to atmospheric CO_2 enrichment. Crop Sci 25:975–981

Blasing TJ (1985) Background: carbon cycle, climate and vegetation response. In: White MR (ed) Characterization of information requirements for studies of CO_2 effects: water resources, agricultures, fisheries, forests and human health. DOE/ER-0236. United States Department of Energy, Washington, DC, pp 10–22

Brown KR (1991) Carbon dioxide enrichment accelerates the decline in nutrient status and relative growth rate of *Populatus tremulioides* (michx.) seedlings. Tree Physiol 8(2):161–173

Den Driessche V (1991) New root growth of Douglas-fir-seedlings at low carbon dioxide concentration. Tree Physiol 8(3):289–295

LaMarche VC Jr, Graybill DA, Fritts HC, Rose MR (1984) Increasing atmospheric carbon dioxide: tree ring evidence for growth enhancement in natural vegetation. Science 225:1019–1021

Landsberg JJ, Kaufman MR, Binkley D, Isebrands J, Jarvis PG (1991) Evaluating progress towards closed forest models based on fluxes of carbon, water and nutrient. In: Kaufmann MR, Landsberg JJ (eds) Advancing towards closed models of forest ecosystems. Herbon Publishing, Victoria. Tree Physiol 9:1–15

Luxmoore RJ (1981) CO_2 and photosynthesis. Bioscience 31:626

Luxmoore RJ (1991) A source sink framework for coupling water, carbon, and nutrient dynamics of vegetation. In: Kaufmann MR, Landsberg JJ (eds) Advancing towards closed models of forest ecosystems. Heron Publishing, Victoria. A special volume of Tree Physiol 9:267–280

Maclaren P (1991) Forestry and the 'greenhouse effect'. N Z For 136(3):23–25

Oechel WC, Strain BR (1985) Native species responses to increased atmospheric carbon dioxide concentration. In: Strain BR, Cure JD (eds) Direct effects of increasing carbon dioxide on vegetation. US Department of Energy, Washington, DC, pp 117–154

Paez A, Hellmers H, Strain BR (1984) CO_2 enrichment and water interaction on growth of two tomato cultivators. J Agric Sci 102:687–693

Poorter H, Pot S, Lambers H (1988) The effect of an elevated atmospheric CO_2 concentration on growth photosynthesis and respiration of *Plantago major*. Physiol Plant 73:553–559

Rastetter EB, Ryan MG, Shaver GR, Melillo JM, Nadelhoffer KJ, Hobbie JE, Aber JD (1991) A general biogeochemical model describing the response of carbon and nitrogen cycles in terrstrial ecosystems to changes in carbon-di-oxide, climate and nitrogen deposition. In: Kaufmann MR, Landsberg JJ (eds) Advancing towards closed models of forest ecosystems. Heron Publishing, Victoria. A special volume of Tree Physiol 9:101–126

Richmond KA, Krag S, Boussibu S (1982) Effects of bicarbonate and CO_2 on the competition between Chlorella vulgaris and Spirulina platensis. Plant Cell Physiol 74:233–238

Rovira AD (1969) Plant root exudeates. Bot Rev 35:35–57

Sasek TW, DeLucia EH, Strain BR (1985) Reversibility of photosynthetic inhibition in cotton after long term exposure to elevated CO_2 concentrations. Plant Physiol 78:619–622

Sharkey TD, Vandeveer PJ (1989) Stromal phosphate concentration is low during feed-back limited photosynthesis. Plant Physiol 91:679–684

Sionit N, Strain BR, Helmers H, Riechers GH, Jaegar CH (1985) Longterm atmospheric CO_2 enrichment affects on the growth and the development of *liquidambar styraciflua* and *Pinus taeda* seedlings. Can J For Res 15:468–471

Solmon AM, West DC (1985) Potential responses of forests to CO_2 – induced climate change. In: White MR (ed) Characterization of informal requirements for studies of CO_2 effects: water resources, agricultures, fisheries, forests and human health. DOE/ER-0236, DOE/ER-0236. United States Department of Energy, Washington, DC, pp 146–169

Strain BR (1985) Background on the response of vegetation to atmospheric carbon dioxide enrichment. In: Strain BR, Cure JD (eds) Direct effects of increasing carbon dioxide on vegetation. US Department of Energy, Washington, DC, pp 1–10

Tolley LC, Strain BR (1984) Effects of CO2 enrichment on growth of liquidambar stryaciflua and P. taeda seedlings under different irradiance levels. Can J For Res 14:343–350

White MR (1985) Objectives of the current study of indirect effects. In: White MR (ed) Characterization of information requirements for studies of CO_2 effects: water resources, agricultures, fisheries, forests and human health. DOE/ER-0236. United States Department of Energy, Washington, DC, pp 2–8

Woodrow IE, Berry JK (1988) Enzymatic regulation of photosynthetic CO_2 fixation in C_3 plants. Ann Rev Plant Physiol Plant Mol Biol 39:533–594

Chapter 5
Role of Vegetation in Reducing CO_2

There are four principal reservoirs of regions of the earth through which carbon flows systematically: the atmosphere, the terrestrial biosphere, the oceans, and the geosphere, which include all forms of fossil. Carbon can be transferred between reservoirs and forms a network called the carbon cycle. The burning of fossil fuel, for example, currently generates an amount of CO_2 equivalent to 5 gigatons (10^{15} g = 1 Gt) of carbon annually. This carbon represents an additional transfer from the geosphere to the atmosphere relative to the balanced natural exchange of about 100 Gt C annually between the atmosphere and each of the remaining major carbon reservoirs resulting in a net increase of CO_2 concentration in the atmosphere (Blasing 1985). Similarly, when a forest absorbs CO_2, deforestation, especially when the forest is burnt, adds considerable quantities of CO_2 to the atmosphere. CO_2 is also released when the forest vegetation decays and through the drowning forest. For example, a single dam, reservoir of the Balbina dam in the Brazilian Amazon, will yield as much CO_2 as would be produced by a coal-fired plant with a similar generating capacity operating for 100 years (Anon 1991).

Thus, it is obvious that there is emission as well as absorption among these major sources. The biomass of a vegetation stand is the end result of the process of carbon uptake, allocation, and loss over the life of the stand. The key ecosystem process involving carbon within a stand includes photosynthesis, respiration, translocation, allocation, storage, fine root turnover, decomposition, herbivory, and shedding. The quality of these processes depends on the age of the plants and the environmental factors. The balance of all these processes is usually treated as carbon sequester. Hence, the role vegetation plays in reducing CO_2 is difficult to describe.

5.1 The Vegetation Carbon Balance

Over periods of a season or a year or more, some of the carbon fixed by a stand may be allocated permanently to stems, large roots, or branches, and some may be fixed in, and then lost from, transient structures such as foliage, shoots, or fine roots.

M. Ali, *Climate Change Impacts on Plant Biomass Growth*,
DOI 10.1007/978-94-007-5370-9_5,
© Springer Science+Business Media Dordrecht 2013

The greatest uncertainties relating to closure of the carbon balance of stands concern carbon allocation to, and dynamics in, these transient structures. However, empirical information on biomass components indicates that they are relatively stable in many cases (Landsberg 1986), so that biomass ratios above-ground components of plants can be used to provide first-order approximations to stand growth patterns. There is so much useful empirical information about foliage and root dynamics. Therefore, the calculation of the stand carbon balance over periods of seasons to years, for stands for which the initial structure and conditions are specified, can be closed within the range of spatial variation encountered in the field (Landsberg et al. 1991). Lack of knowledge of mechanisms governing the dynamic processes of carbon allocation, and uncertainties about the dynamics of transient structures, makes closure on these time scales less reliable than on shorter time scales. Thus, the priority areas for the future research on the long-term carbon balance to achieve process-based models are carbon allocation, below-ground processes (root dynamics), and foliage dynamics. We must also improve our empirical knowledge of respiration and decomposition, our knowledge of interactions with water and nutrients, and our ability to represent these processes in the models.

Over short periods, in vegetation, the main loss of carbon is through respiration. Respiration measurement in the field is difficult, and the ability to account for these losses is still limited. Nevertheless, the errors in measurement may not be large enough to impose critical limitation on carbon balance studies within forest stands. The indications are that, over short periods, one may achieve closure in carbon balance calculations for stands of specific structure and physiological characteristics, with errors that are less than the range of variation within stands.

A major question is how terrestrial ecosystems, especially vegetation, will respond to a regional and global increase in atmosphere CO_2 concentrations, temperature, and N inputs. It is of particular interest how these changes will affect terrestrial C storage; that is, will terrestrial ecosystems store more C and slow the accumulation of atmospheric CO_2 or release C to increase the rate of atmospheric CO_2 accumulated? To answer these questions, it is important to understand what controls the dynamics of C within terrestrial ecosystems and the exchanges of these C with the atmosphere.

5.2   Control Dynamics of C by N

Linkages between the C and N cycles regulate the amount, distribution, and turnover rates of C within many terrestrial ecosystems. Plant growth rates and tissue N contents, constrained by the availability of N and nutrition, strongly influence the response of photosynthesis to increased atmospheric CO_2. Similarly, decomposition of C materials and N mineralization are constrained by the need for soil microbes to maintain their own C/N balance. These C/N interactions may constrain the biogeochemical responses of terrestrial ecosystems to changes in atmospheric CO_2 concentration, climate, and N inputs.

Very subtle shifts in the C–N metabolism of terrestrial ecosystems can have a profound impact on the amount of C stored. For example, the C to N (C/N) ratios of terrestrial vegetation are typically 40 or higher if there are substantial amounts of woody tissue. Soils, on the other hand, typically have a C/N ratio 20 or less (Rastetter et al. 1991). Thus, each unit of N in the soil holds half as much C as does the equivalent amount of N in vegetation. If decomposition of soil organic matter were to increase, and the N thereby released from soils (at a C/N of 20) were to be taken up into the plant biomass (at a C/N of 40), then the amount of C stored in the ecosystem as a whole would increase by 20 g for each gram of N released from the soil. Alternatively if the increased N released in the decomposition were leached out of the ecosystem, there would be a loss of 20 g of carbon per gram N released. Similar scenarios can be imagined with very subtle shifts between woody tissue (at a C/N of 300 or higher) and nonwoody tissue (at a C/N near 20).

Carbon is distributed differently in different ecosystems, for example, 95% carbon in tundra, and only 50% of the total carbon in forests is stored in soils (Rastetter et al. 1991). Soils of both ecosystems are inceptisols (USDA term) with organic horizons containing large stocks of C and N overlaying mineral horizons with much lower organic content.

Differences between two ecosystems arise partly because of differences in the initial distribution of C and N between vegetation and soil organic matter. Most of the N in both ecosystems is in the soil (about 99% in tundra and 90% in hardwood forests). The soil has a relatively low C/N ratio, and the uptake of N by plants, which have higher C/N ratios, represents very large potentials for increased whole ecosystem carbon storage even if there is no change in total ecosystem N storage. However, differences between tundra and hardwood forests in the C/N ratios of these vegetation and soils may also influence this potential. There is far more C per unit N in hardwood vegetation than in tundra vegetation. Thus, the difference between the C/N ratios of vegetation and soils is greater ($146.3 = 159.5 - 13.2$; Table 4.5) in hardwood forests than (only $24.1 = 41.6 - 17.5$; Table 4.5) in tundra. So for each gram of N transferred from soil to vegetation, there will be about six times ($146.3/24.1$) as much net C storage in hardwood forests as in tundra. Transfer of N between vegetation and soil do not always result in the changes in C storage that would be predicted from their respective C/N ratios. For example, when CO_2 is increased in the tundra, there may be a little shift of N from vegetation to soil. This in itself will tend to decrease the amount of C stored in the ecosystem because of the lower C/N ratios of the soil. However, this effect is usually counteracted by a corresponding increase in the C/N ratios of the vegetation, mostly resulting from an increase in the stem tissue (Rastetter et al. 1991).

In general, linkages between C and N often act to constrain long-term responses of disturbances in both forest and tundra ecosystems. Because the amount of N imported from outside these ecosystems is very small relative to the N recycled internally, the ability of both the ecosystems to respond to perturbation is limited to their ability to redistribute existing internal nitrogen among the major organic matter pools and their ability to adjust the C/N ratios within these pools. The potential for sequestering C within the ecosystem may, therefore, be limited by the amount of N that can be distributed from components with low C/N ratios (soils and leaves) to components of high C/N ratios (wood and stem) and by constraints on changes of C/N ratios of these components. If the assumed incidence of increased temperature does happen due to climate change, then N turnover rate may increase, and a consequent shift in N from soil to vegetation may take place, thus fixing more carbon. When both temperature and CO_2 increase together, the long-term responses may be dominated by the temperature effect on N cycling. Further researches would be necessary to confirming the issues related to temperature.

5.3 Growth Enhancement and Reduction of Carbon

The amount of carbon reduced by vegetation can be determined by the growth nature of the vegetation. In order to model the response of the terrestrial biota to increase level of atmospheric CO_2 when formulating a global cycle model, Gates (1985) mentioned a biotic factor β defined by Bacastow and Keeling (1973) as the fractional change in the atmospheric CO_2 concentration and may be written as $\beta = (\Delta NPP/NPP_0)/(\Delta Ca/Ca_0)$ where NPP_0 is the initial value of the net primary productivity for an initial value of the atmospheric CO_2 concentration denoted by Ca_0. Whether β value is used or not, the determination of growth rate is important in determining the change of productivity.

In natural conditions, the change in growth may be identified by examining the growth rings. LaMarche et al. (1984) reported the increased growth rates in the annual rings of subalpine conifers in the Western United States since the mid-nineteenth century. Those growth rates were shown to be consistent with the magnitude of global trends in the atmospheric CO_2 concentration. LaMarche et al. (1984) believe that subalpine vegetation generally, and upper tree line conifers in particular, could be exhibiting enhanced growth as a direct response to increasing concentration of atmospheric CO_2. In fact, the authors showed that the increase in the average growth rate of high-altitude bristlecone pine and limber pine is constant with a 26% increase in the atmospheric CO_2 concentration from 1850 to 1950 (from 270 to 340 ppm).

Olson et al. (2001) and Ajtay et al. (1979) reported that the total plant carbon in the world is about 559 Gt (gigaton), not counting the ocean which contributes about 3 Gt. The NPP of the terrestrial ecosystems is about 60 Gt per year. Gifford (1980) calculated the annual increment of carbon stored in wood and soil organic matter (from the biosphere measurement in 1979 when atmospheric CO_2 concentration was assumed as 316 ppm) to be 1.65 Gt per year. Of this amount, 65% is stored as wood and 35% as soil organic matter. The total soil organic pool of carbon in the world was estimated by Schlesinger (1984) to be about 1,500 Gt. Therefore, any increased store estimate above as soil carbon is too small a change to be detectable. Of the total carbon estimated by Gifford (1980), the greatest amount of storage is probably by the tropical savannahs and grasslands (0.43 Gt/year about 26% of the total), the next greatest by tropical forests (0.32 Gt/year about 19% of the total) followed by temperate forests (0.20 Gt/year) and temperate grasslands (0.16 Gt/year). Lakes, bogs, and streams are estimated to store about 0.21 Gt/year of carbon, when β value was taken as 0.60 (Gates 1985).

Among these storages, the amount of carbon stored in wood (1.07 Gt/year; Gates 1985) is a temporary storage because eventually this wood will decompose. Gates has mentioned that if β value is taken as 0.60, wood decay will equal to or exceed the rate of sequestering, perhaps inside 100 years. If the β is taken as 0.37, about 1.0 Gt/year of carbon would now be stored by the biosphere. Recalling β as the ratio between the ratios of changes in ecosystem productivity to the ratios of CO_2 concentration in atmosphere, a lower-value β dictates lower change of NPP. This can be summarized in a way that fixation of CO_2 through green vegetation will start contributing to emission in the long-term perspective. If these happen to be true, then uncontrolled C trading to reduce carbon emission may not be in the long run a sustainable solution to climate change.

5.4 Human Impact on C Dynamics

In the introductory section of this chapter, it was mentioned that one of the misbalancing activities in natural transfer of C among four major sources is human activity. Human activities may produce imbalances in the C cycle either through direct impact on C emission processes or indirect impact through manipulating N cycle or other C-sequestering processes. Industrial and agricultural activities are responsible for increased input of N in many mid-latitude ecosystems. An estimate by Landsberg et al. (1991) shows dry plus wet deposition of N to all temperate ecosystems could be over 14 Tg (10^{12} g) per year. Hence, there is a possibility of collecting and storing excess CO_2 in the form of trees, their wood products, and in the forest soils.

When a forest is growing, it acts as a net sink for CO_2, whereas mature or old growth forests are assumed to maintain CO_2 equilibrium with the atmosphere. If forests are harvested and planted periodically, carbon is fixed in living trees during regrowth and put into storage in various wood products, such as building materials, furniture, and various detritus and soil carbon pools which subsequently decay on different time scales. Thus, repeated harvesting accumulates a stock of carbon only over a finite period, after which an equilibrium is reached where further C fixation by forest regrowth would be balanced by the decay products of wood and soils back to CO_2. If this balance could be maintained at the level of the old growth forests by harvesting and reforestation, then periodic felling and replanting may be better, in respect of the CO_2 balance, than reserving the forests aside permanently. If the impact of human activities on vegetation is considered, the above statement reveals that controlled and managed human activity may not necessarily hamper the carbon cycle in the long term. Therefore, nature of human operation on vegetation, that is, management of forest, plays a critical role in CO_2 balance. To have some ideas about these issues, we draw the attention of readers to the next chapter on how a managed forest plays a role in C balance.

References

Ajtay GL, Ketner P, Duvigneaud P (1979) Terrestrial primary production and phytomass. In: Bolin B, Degens ET, Kempe S, Ketner P (eds) The global carbon cycle. Wiley, New York, pp 129–171

Anon (1991) Rainforest destruction: causes effects and false solutions. World Rainforest Movement, Malaysia, Jutaprint

Bacastow R, Keeling CD (1973) Atmospheric carbon dioxide and radiocarbon in the natural carbon cycle: II. Changes from A.D. 1700 to 2070 as deduced from a geochemical model. In: Woodwell GM, Pecan EV (eds) Carbon and the biosphere. CONF-72051, Technical Information Center, Office of Information Services, United States Atomic Energy Commission, pp 86–135

Blasing TJ (1985) Background: carbon cycle, climate and vegetation response. In: White MR (ed) Characterization of information requirements for studies of CO_2 effects: water resources, agricultures, fisheries, forests and human health, DOE/ER-0236. United States Department of Energy, Washington, DC, pp 10–22

Gates DM (1985) Global biosphere response to increasing atmospheric carbon dioxide concentration. In: Strain BR, Cure JD (eds) Direct effects of increasing carbon dioxide on vegetation. US Department of Energy, Washington, DC, pp 171–184

Gifford RM (1980) Carbon storage by the biosphere. In: Pearman GI (ed) Carbon dioxide and climate: Australian research. Australian Academy of Sciences, Canberra, pp 167–181

LaMarche VC Jr, Graybill DA, Fritts HC, Rose MR (1984) Increasing atmospheric carbon dioxide: tree ring evidence for growth enhancement in natural vegetation. Science 225:1019–1021

Landsberg JJ (1986) Physiological ecology of forest production. Academic, London

Landsberg JJ, Kaufman MR, Binkley D, Isebrands J, Jarvis PG (1991) Evaluating progress towards closed forest models based on fluxes of carbon, water and nutrient. In: Kaufmann MR, Landsberg JJ (eds) Advancing towards closed models of forest ecosystems. Tree physiology. Heron Publishing, Victoria, 9:1–15

Olson JS, Watts JA, Allison LJ (2001) Major world ecosystem complexes ranked by carbon in live vegetation: a database. NDP-017, carbon dioxide Information Center, Oak Ridge National Laboratory, Oak Ridge, Tennessee

Rastetter EB, Ryan MG, Shaver GR, Melillo JM, Nadelhoffer KJ, Hobbie JE, Aber JD (1991) A general biogeochemical model describing the response of carbon and nitrogen cycles in terrstrial ecosystems to changes in carbon-di-oxide, climate and nitrogen deposition. In: Kaufmann MR, Landsberg JJ (eds) Advancing towards closed models of forest ecosystems. A special volume of *tree physiology*. Heron Publishing, Victoria 9:101–126

Schlesinger WH (1984) Soil organic matter: a source of atmospheric CO_2. In: Woodwell GM (ed) The role of terrestrial vegetation in the global carbon cycle: measurement by remote sensing. Wiley, London, pp 111–127

Chapter 6
The Carbon Balance in the Managed Forest

The area of the world's managed forest is increasing at a rapid rate to meet the demand of a growing population with which natural forests cannot cope. An estimated increase of global population and the demand for wood is provided in Table 6.1 from which it is clear that managed forests are important in meeting the future demand of timber.

Sutton (1991) estimated that there could be around 100 million ha of plantation forest around the world. Of these, nearly 14 million ha are fast growing (defined as having a mean annual increment of 24 m^3/ha/year). While slow-growing trees are mostly used for construction, fast-growing species have diverse uses. Even in many cases, use of hardwood species has been supplemented by technological development of softwood use to meet the increased demand of a growing population. Hence, raising a fast-growing plantation rather than a slow-growing plantation is increasing in popularity.

The global distribution of fast-growing plantations and their corresponding uses are shown in Table 6.2 and Fig. 6.1. These figures may give some appreciation of the extent to which the fast-growing plantations are responsible for carbon sequestering. The summary presentation in the form of different end uses is provided because C balances for fast-growing forests depend on how the forests are managed and what the forest products are used for. The uses of fast-growing trees in the form of pulp, sawlogs, and pruned areas cause semipermanent sequestration of C. However, in developing countries, fast-growing species are used for fuel which emits the carbon back to atmosphere.

6.1 Role of Forest Trees and Their Products

For a managed forest, the average stock of C in living trees and soil can be as little as 30 and 70%, respectively, of that for old-growth forests (Covington 1981; Cooper 1983). The carbon balance between the atmosphere and managed

Table 6.1 Wood demand (in billion m³) and population (in billion) up to 2010

	Production project		
	1989	2010	Increase
Industrial wood	1.68	2.58	0.90
Fuelwood	1.78	2.38	0.60
All wood	3.46	5.07	1.50
Population	5.10	6.80	1.90

Source: Sutton (1991) and Internet sources

Table 6.2 Estimated fast-growing[a] plantations by region, conifer/ hardwood, and management 1990 (areas in million ha)

	Conifers			Hardwood	
Region	Pulp[b]	Sawlogs	Total	Pulp	Total
Africa	0.4	0.6 (0.5)	1.0	0.9	1.9
Asia (Central and South)	0.4		0.4	0.25	0.65
America	2.0	1.9 (0.3)	3.9	3.9	7.8
Europe	0.05	0.2	0.25	0.8	1.05
Oceania	0.2	1.9 (1.2)	2.1	0.1	2.2
Totals	3.05	4.6 (2.0)	7.65	5.95	13.6

Source: Sutton (1991)

[a]Fast growing is defined as having mean annual increment greater than 14 m³/ha/year

[b]Pulp includes all non-saw log objectives (other than fuel)( ) Volumes in brackets are estimates of areas pruned

Fig. 6.1 Fast-growing plantations – 1990 (in million ha) based on Table 6.2 (Source: Sutton 1991)

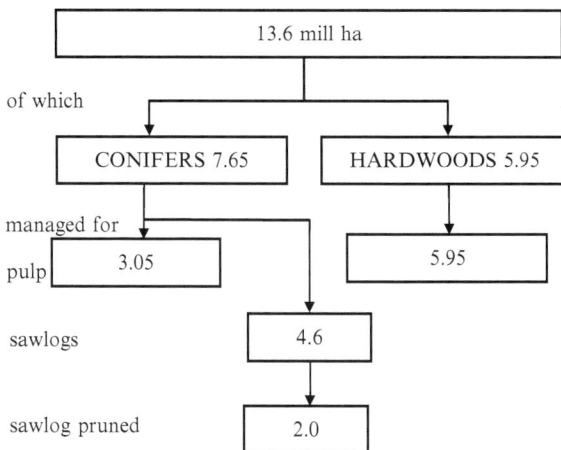

forests depends critically on the life span of the wood products, as well as on the frequency and intensity of harvest. For a short-lived wood product, such as paper, the stock of C accumulated may be too small to offset the deficit in trees and soil. A significant amount of C could be accumulated by fast-growing tree species which were harvested and used for reasonably long-lived products, when compared with more traditional species with current products (Dewar 1991). Harmon et al. (1990)

concluded from a simulation of Douglas-fir and hemlock ecosystems, typically of the Pacific Northwest, that the reduction in stored carbon on conversion of old-growth forest to a 60-year rotation forest could be offset only by approximately doubling the life span of wooden buildings.

A more general approach was taken by Dewar (1990), who used a simple analytical model of carbon in trees and wood products to show that carbon accumulation alone in wood products offset the loss of carbon incurred by felling trees, only if the lifetime of the wood product (Dp) exceeds the stand age at a maximum mean annual increment (T_m). In addition, the equilibrium wood product carbon pool is proportional to Dp/T_m and is large for either long-lived products (low decay) or short-rotation forests (high input rate). Hence, this ratio of time scale is critical for the model. However, the Dewar model did not take into consideration the C pool of the forest floor, which is sometimes the largest pool in the temperate forests.

6.1.1   The Forest Floor

Useful information regarding forest floor biomass has been obtained from the study of northern hardwood forests in the USA. The forest floor organic matter undergoes an initial period of decline after harvesting, followed by a period of recovery toward preharvest level. This response is attributed to a lag in the recovery of litter input by the next crop, particularly the less decomposable woody litter, and the stimulation of soil decomposition by increase in water, temperature, and nutrient availability in the first few years after harvest (Dewar 1991). However, before declining, forest floor biomass may increase during the initial period after harvest as a result of slash debris left on site. By simulating a succession of rotations, Aber et al. (1978) found that the rotation period had a much greater effect on forest floor biomass than harvesting intensity, with short rotation and complete harvesting causing the greatest reduction in the forest floor biomass.

The result obtained by Dewar (1990) can be extended by incorporating a soil sub-model and stating the effects of harvesting on inputs of litter and slash debris. Within an extended model, Luxmoore (1991) has identified a dimensionless parameter "α" that controls the degree to which the C accumulation in harvested biomass (wood products and debris) offsets the loss of C in the living trees and soils. Parameter "α" is the generalization of the ratio Dp/T_m in Dewar (1990). For "α" less than critical value α_c, conversion of old-growth forest to managed forest releases carbon to the atmosphere. For a range of forest types, it has been found that the value of α_c is to the order of 1 (Dewar 1991). The sub-model below can be considered for estimating a rotation for single-species forest.

In a forest ecosystem, C pool is regulated by many factors that include mineralization, vegetation, and fluctuation of physical climate. Some of those factors are taken into consideration in the following paragraphs.

6.1.2   The Tree C Pool

The tree C pool represents C stored in the living parts and is a sink for CO_2 while the tree is growing. The ecophysiological process that regulates the growth process of trees plays an essential role in the recycling of C. Carbon storage represents, by living trees at a time (t) after planting, the net result of photosynthesis minus respiration and mortality up to that time. The quantity of stored C in time t ($gc(t)$) is generally represented by the woody parts (stem, branch) of the trees. Hence, C content can be described by a curve similar to the growth curve (sigmoid) approaching a constant value in the old-growth trees (let g_m, Mg C ha^{-1}). Foliage and fine roots follow a different pattern generally reaching closure well before the old-growth stage. However, the contribution of leaves and fine roots to carbon storage is relatively small, usually reported as 4–6% only (Dewar 1991). Hence, carbon stored in a forest can be estimated from the growth function of that forest if the fractional attribute is obtained. In a simple statement, the fractional statement can be described as $y(t) = g_c(t)/g$, from which it is clear that the fraction $y(t)$ is dependent on time or age of the forest trees. The fraction may become a little complex if we consider the differences in C content of the sapwood and heartwood because often the volume of sapwood varies with the age of trees. Thereby, instead of using a growth model, a tree C pool model, formulated from differential contents of C in different component of trees multiplied by the mass/volume proportion of different components of plants at a particular age, can be used for making a harvesting plan of forest crops.

6.1.3   Impact of Harvesting on Carbon Pool

After each time period t (rotation), a proportion of the C is removed from the forest. The quantity of carbon removed depends on the harvesting intensity. Harvesting intensity means the proportion of material taken out of the forest leaving the debris, slash, and tops behind. Total removal of all plant material in harvesting is unlikely, except, perhaps in densely populated areas where firewood is scarce. Typically, 10–20% of a stem is left on the site as broken pieces, stools, and branches (Dewar 1991). With time, the leftover slash will decay, and the carbon pool on the forest floor will decline. On the other hand, if the area is replanted/regenerated and the site quality does not change after harvest, new accumulation of carbon can be described by the same growth function $g_c(t)$. If the degeneration model for C loss from the floor can be derived, the total carbon content of a forest at a particular stage can be derived by algebraic addition of these two models' function. Both model functions are dependent on the site quality. If the traditional site quality expression for a particular species could be amalgamated with the expression of these functions, carbon accumulation can be determined at a particular time. If such a function could be derived, an environmentally sustainable silvicultural system could be prescribed instead of the present sustainable system based only on yield. In such a system, consideration of appropriate habitat measures can also be incorporated. In addition, the senescence property of that species at that particular site may need to be considered with time.

6.1.4   Carbon in Wood Products

The product carbon pool C_p represents carbon stored in different wood products after harvesting that has not yet decayed back to CO_2. The size of the pool depends on the product type. On the other hand, the type of product for which the wood would be used mostly depends on the diameter of the tree. However, a tree can be used for different types of products P_x (when $x = 1, 2, 3, \ldots x$) having different time spans of carbon retention. Let us denote by $r_j(t)$ the fraction of carbon retained by the product j ($j = P_1, P_2, P_3, \ldots P_x$) at time t after felling. If a curve can be drawn based on the amount of carbon retained against time, this may be called the carbon retention curve. However, the C retention curve will also include factors from technological marvel, socioeconomic, and climatic variables and how they are used. The dynamics of the product C pool depend on this carbon retention curve and the wood product's lifetime integral of the weighted mean of these curves. If U_j denotes the fraction of removed material used for product j, then following Dewar (1991) we can write:

$$C_p = \int_0^t \sum_{j=1}^p U_j r_j(t) dt; \text{ where } C_p \text{ is the dynamics of carbon pool during}$$

product lifetime.

6.2   The Soil Carbon Cycle

During the lifetime of a forest, the floor continuously receives carbon from the trees in the form of dead litter and roots which subsequently decompose in the process of soil formation. There is also a discrete input at the end of each rotation in the form of debris left on the site, the amount of which depends on the growth stage of the harvesting, and its intensity. Hence, a distinction could be made between decomposed material incorporated in the soil and undecomposed litter.

6.2.1   Litter and Slash Input

Foliage (L_1) and fine-root (L_f) litter production rates (Mg C ha^{-1} year^{-1}) can be modeled as a function of time after planting by an asymptotic approach to their old-growth values β_1 and β_f (Mg C ha^{-1} year^{-1}), where old-growth values β_1 and β_f are constant for a particular type of forest. In that case, a dimensionless nonwoody litter production rate can be used to estimate the litter input [$Y(t)$]. If it is asymptotic, the relation could be described as $Y(t) = 1 - e^{-2.3t/tc}$, where t_c (year) is the time taken to reach 90% old-growth productions (Dewar 1991). Hence, the productions of different litter in time t may be represented as

$$L_1(t) = \beta_1 Y(t)$$

$$L_f(t) = \beta_f Y(t)$$

$$L_b(t) = \beta_b Y(t)$$

$$L_s(t) = \beta_s Y(t)$$

$$L_w(t) = \beta_w Y(t)$$

where suffixes l, f, b, s, w represent the leaf, fine root, branch, stem, and woody-root debris, respectively, with an assumption that for each component mortality is proportional to standing biomass. The continuous input of litter is added by discrete input from stem, branch, and woody-root debris left on site at harvest as specified by the fraction h_b, h_s, and h_w resulting in a step increase in the corresponding litter pools at the end of each rotation:

$C_i \rightarrow L_i(t) + h_i g_c(t)$; $i = 1, f, b, s, w$ and $g_c(t)$ is the proportion of C content

   at the time of harvest.

6.2.2   Decomposition

For each component of litter, the residual products are transformed into soil. Let us take fractional decomposition rates k_i per year for litter compartments $i = 1, f, b, s$, and k_m per year for soil organic matter. In addition, we define μ_i for $i = 1, f, b, s$ as the fraction of litter compartment i that becomes soil, the rest being lost as CO_2.

6.2.2.1   Dynamic Equations for the Soil Sub-model

By combining the litter input and decomposition, the dynamic behavior of the soil sub-model within the interval between the two consecutive harvests can be described by the following first-order differential equations:

$dC_i/dt = L_i(t) - k_i C_i$,   where $i = 1, f, b, s, w$ and L_i is the rate of litter production.

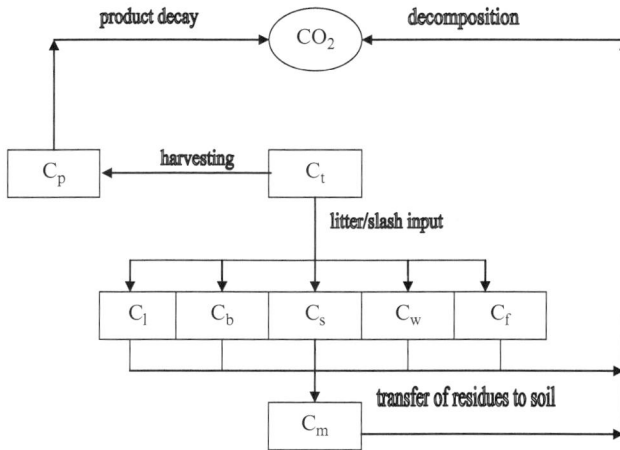

Fig. 6.2 Schematic representation of C storage model. *Boxes* represent the pools of C in living trees (t), wood product (p), foliage litter (l), branch litter (b), stem litter (s), woody-root litter (w), fine-root litter (f), and in soil organic matter (m). *Arrows* represent carbon fluxes into and out of each pool. The *circular box* represents the source sink of CO_2 (Source: Dewar 1991)

$dC_m/dt = \Sigma \mu_i k_i C_i - k_m C_m$ where $i = l, f, b, s, w$ and μ is the constant that determines the amount of decomposed portion that gets into soil under certain climatic/seasonal conditions.

If $dC_m/dt = (1 - g_c(t))$, then we can say that there is no carbon storage taking place in the system; if $(dC_m/dt) > (1 - g_c(t))$, the system is fixing carbon; and if $(dC_m/dt) < (1 - g_c(t))$, the system is losing carbon. A system might lose carbon when the forest is clear felled or thinned very heavily, that is, the proportion of carbon left $(g_c(t))$ is small.

In the above few paragraphs, we have tried to determine the dynamics of carbon distribution of all components of Fig. 6.2. However, the issues may not be so simple. Where the CO_2 concentration in the atmosphere increases, there will be changes in the growth and litter fall processes. Whether these changes will be proportional to all compartments is as yet unknown, but it seems from these equations that, on a long-term basis, CO_2 emitted from deforestation will be balanced by new growth, that is, if replanted. But the problem is to deal with the CO_2 emitted from other sources resulting in continuous increase in its concentration in the atmosphere. Where such a gradual increase is not checked, the soil nutrient pool may be exhausted by the increase of the aerial fertilization effect which may reduce further biomass growth if additional fertilizer is not added. Under this situation, reduction in biomass growth may end in a further cumulative increase in CO_2 concentration.

If we look at the figure above, we can say that after a number of rotations, an equilibrium condition of storage of average C content may be achieved in a managed forest, but the actual C content will vary around this average level. Figure 6.3 reveals that the average level of C in a plantation can be increased or decreased

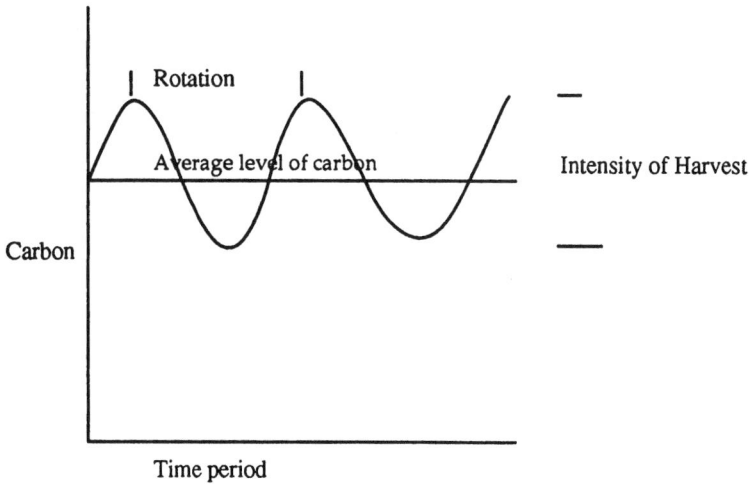

Fig. 6.3 Estimated level of carbon variation in a sustainable managed forest

by manipulating the rotation period and harvesting intensity of that plantation. However, for a particular ecosystem, a minimum level of rotation or maximum level of harvesting intensity could be prescribed beyond which the fluctuation of C content after each rotation may reach a critical value (toward zero) at which the total ecosystem could be changed and the system may start in a new direction. Prescription of such rotation system will depend on the species, climate, and the rate of decomposition and intensity of harvesting.

Biomass growth activity and CO_2 gain may also vary from plant to plant and from their association with other plants. Reid et al. (1983) found higher photosynthetic rates in needles of ectomycorrhizal Loblolly pine (*P. taeda*) than in nonmycorrhizal plants. Paul and Kucey (1981) observed greater carbon assimilation in plants was enhanced by the presence of nitrogen-fixing bacteria. Individual leaves of oak seedlings undergo cyclic changes in photosynthetic rate reflecting the intermittent growth of the shoot which occurs in a series of flashes (Hanson et al. 1988). Leaves of fruiting trees have a higher photosynthetic rate than leaves of nonfruiting trees during the period of rapid fruit growth (Fujii and Kennedy 1985). From these findings, it can be presumed that a good (C) sink could be developed through village forestry or agroforestry if genetically improved fruit trees are incorporated in plantations. Hence, climate change may accentuate other growth which is not considered as woody biomass growth in a true sense.

Under certain conditions, non-woody biomass growth can lead to greater sink activity. Downtown et al. (1987) reported that CO_2 enrichment during fruit development may increase fruit yield Dickson's (1989) findings say that during the spring growth flush both stored and recently fixed carbohydrates contribute to new shoot growth. Early wood development in ring porous trees begins before shoot growth and depends on stored reserves, but late-season diameter growth depends entirely on current photosynthesis. Thus, current photosynthesis frequently goes directly to support non-woody biomass growth.

In many cases, photosynthates may be stored and used later in biomass growth. In many temperate forest species, shoot growth is determined by the number of performed (photosynthetically active) stem units and leaf primordia, expansion of which may depend largely or entirely on stored reserves. In such an activity, stored minerals play an active role. Stem thickening and root growth are usually indeterminate, but they can show periodicity when root growth is reduced during periods of shoot growth. In that case, CO_2 used during the growing period can only be transformed to biomass if availability of minerals in the store/environment is adequate.

References

Aber JD, Botkin DB, Melollo JM (1978) Predicting the effects of different harvesting regimes on forest floor dynamics of Northern Hardwoods. Can J For Res 8:306–315

Cooper CF (1983) Carbon storage in managed forests. Can J For Res 13:155–166

Covington WW (1981) Changes in forest floor organic matter and nutrient content following clear cutting in Northern Hardwoods. Ecology 62:41–48

Dewar RC (1990) A model of carbon storage in forests and forest products. Tree Physiol 6:417–428

Dewar RC (1991) Analytic model of carbon storage in the trees, soils, and wood products of managed forests. Tree Physiol 8(3):239–258

Dickson RE (1989) Carbon and Nitrogen allocation in trees. Ann Sci For 46:631–647

Downtown WJS, Grant WJR, Loveys BR (1987) Carbon dioxide enrichment increases yield of Valancia orange. Aust J Plant Physiol 14:493–501

Fujii JA, Kennedy RA (1985) Seasonal changes in the photosynthetic rate in apple trees. Plant Physiol 78:519–524

Hanson PJ, Isebrands JG, Dickson RE, Dickson RK (1988) Ontogenetic patterns of CO_2 exchange of *Quercus rubra* L. leaves during three flushes of shoot growth of median flush leaves. For Sci 34:55–86

Harmon ME, Ferrell WK, Franklin JF (1990) Effects on carbon storage conversion of oldgrowth forests to young forests. Science 247:699–702

Luxmoore RJ (1991) A source sink framework for coupling water, carbon, and nutrient dynamics of vegetation. In: Kaufmann MR, Landsberg JJ (eds) Advancing towards closed models of forest ecosystems. A special volume of *tree physiology*. Heron Publishing, Victoria 9:267–280

Paul EA, Kucey RMN (1981) Carbon flow in plant microbial associations. Science 213:473–474

Reid CPP, Kidd FA, Ekwebelam SA (1983) Nitrogen nutrition, photosynthesis and carbon allocation in ectomycorrhizal pine. Plant Soil 71:415–432

Sutton W (1991) Are we too concerned about wood production? N Z Forest 36(3), New Zealand Institute of Forestry Inc., pp 25–27

Chapter 7
Short- and Long-Term Resolution

From the rotational curve presented in Fig. 6.3, we can understand that estimation of climate change impact may reveal different results on short-term and long-term bases. In practice, increase in the concentration of CO_2 is a slow process, and it may take a long period to predict any change in the biomass growth from the CO_2-induced effects. But both short- and long-term resolutions are important for management purposes. Positive management of climate change effect may be possible by looking at the reduction of greenhouse gas emission which can be a short-term as well as a long-term project or by improving the adaptability of green biomass through genetic manipulation which may be a long-term project. We are considering here the different aspects of short- and long-term resolution of climate change impacts on vegetation.

7.1 Aspects of Source-Sink Activity

In any strategic measures for control of the greenhouse effect, most emphasis should be given to the carbon among the components. The change in the concentration of CO_2 in the atmosphere is the result of imbalance between the sources and sinks. We have mentioned earlier that the sources of CO_2 are mainly fossil fuel/biomass fuel/industry and deforestation and the major sinks are plants and oceans. But estimates of emission of CO_2 from the sources, and inputs to the sink, vary from time to time that can have significant impact on long-term and short-term resolution. Tucker (1990) reported that the magnitude of CO_2 emission from industries is about 5 gigatons (Gt/year). Atmospheric measurement of CO_2 concentration indicates that the accumulation rate is 2.9 Gt/year. It is interesting that the ratio of uptake by the biosphere and oceans over emissions has remained at about 42% despite increase in fossil fuel emissions. The differences in the trends of emission and accumulation in sink estimated by Tucker (1990) have been shown in Table 7.1 along with the balances.

M. Ali, *Climate Change Impacts on Plant Biomass Growth*,
DOI 10.1007/978-94-007-5370-9_7,
© Springer Science+Business Media Dordrecht 2013

Table 7.1 Schematic representation of historical sequence of estimates of principal CO_2 exchanges with the atmosphere (Gt/year) and the increase in the atmospheric burden

	Ecology/forest			Burden		
Year	Equatorial	Temperate	Industry	Ocean	Atmosphere	Missing sink
1980	3	−1	5	−2	3	2
1985	2	−1	5	−2.5	3	0.5
1989	2	−1	5	−1	3	2

−ve figures indicate sink and +ve figures are emission (*Source*: Tucker 1990)

The figures in Table 7.1 show that the variation is mainly associated with the forest source and the ocean sink. Further investigation opens up the question because oceans, as a sink for CO_2, appear incapable of accommodating more than 1 Gt/year of CO_2. The distribution of sources of and sinks for CO_2 is also not uniform. Enting and Mansbridge (1989) suggested that the Northern Hemisphere (NH) oceans, which are smaller in area than the Southern Hemisphere (SH) oceans, are alone incapable of providing the required NH mid-latitude sink for carbon. The detailed study of Tans et al. (1990) indicates that the differences between the partial pressures of CO_2 in surface ocean waters and the atmosphere in the NH are too small for the oceans to be the major sink of CO_2. Thus, it is possible that the terrestrial biosphere may be a comparable or an even larger sink for CO_2 than the oceans. In the long run, therefore, the option remains to control emission in the biosphere. In the biosphere, green biomass remains the potential sink. We have seen from discussion before that in long-term perspective, the emission and sink from green biomass balance each other. If certain actions are not taken – it is unlikely that the reduction of emission is possible in the long run. Let us discuss the prospects here.

7.2   Aspects of Plant Growth

Plants are one of the important terrestrial sinks. From isotopic study, it has been found that land plants selectively fix more ^{12}C than ^{13}C during photosynthesis. No selection occurs as a result of exchange with the ocean mainly because the ocean biota get their carbon from inorganic sources in the ocean; only 1 % of ocean carbon is dissolved CO_2 (temp and amount of dissolved carbon). The terrestrial sink for CO_2 also depends on the ratio of $^{13}C/^{12}C$ in the atmosphere (Tucker 1990). However, they vary from species to species.

Short-term leaf responses of plants to increasing CO_2 may be different depending on the type of carbon fixation pathway present in a particular plant species. Callaway and Currie (1985) stated that plants with a four-carbon (C_4) pathway close their stomata more tightly in the presence of high CO_2 concentrations than do plants with a three-carbon pathway (C_3). Consequently, C_4 plants do not respond as vigorously to the fertilization effects of high CO_2 as do their C_3 counterparts. On the other hand,

C_4 plants need less water to maintain their growth than the C_3 plants. But water-use efficiency (photosynthesis rate/transpiration rate) of both types of plants increases with ambient CO_2 concentrations. This means that the amount of transpiration per unit of leaf area should decrease for both C_3 and C_4 plants. As a result, the short-term effect of less precipitation due to climate change on plant growth will depend on the nature of the plant as well as on the scale of precipitation.

But the long-term impact of CO_2 increase is still not clear. Although scientists are reasonably sure that transpiration rate per unit leaf area will decrease as CO_2 increases, they are not sure about the consequences at the ecosystem level should the total leaf area of the plant community increase, contradicting the leaf-level response. Increasing leaf area will increase canopy interception, reduce soil surface evaporation, and decrease incident radiation in lower canopies for photosynthesis, and as a result, total canopy-level transpiration may be increased among other ecosystem feedback. In addition, changes in community species composition and geographical range may occur. As the ambient concentration of CO_2 increases, C_4 plants may be able to extend their range in the drier conditions, whereas C_3 plants may be favored where the soil moisture is not limiting. Hence, the nature of the buildup of vegetation canopy will control the soil moisture and the amount of solar radiation reaching the ground and may affect the total impact.

Biomass growth due to increase in the concentration of atmospheric CO_2 will partly depend on the adaptability of the species to the changes in C concentration and the subsequent climatic changes. The adaptability among the perennial and annual plant species may vary depending on the rate of change in the climate and the CO_2 concentration. Kimball (1985) predicted that plants may adapt themselves to the higher concentration of CO_2 in the atmosphere through changing the stomatal aperture. If the next generation is not grown under the same or higher concentration, then the change in stomatal opening may no longer remain and hence will not be inherited. But if the succeeding generation has the same changes in the stomatal opening, then this would show that the plant had adapted to the changes. If we take 70 years as the period within which doubling of CO_2 concentration will occur, by this time, annual species will have many generations to adapt to the changes, whereas individuals of long-lived perennials will only have a part of their life cycle to become adjusted or acclimatized. Hence, in the long run, annuals may invade the ecosystem, and such invasion most likely will result in the gradual exclusion of perennials from the system. Consequently, in the unmanaged biosphere, natural selection will determine which plants are best adapted or most fitted for the future high-CO_2 world. The plants with characteristics that ensure greater survival and reproductive advantages in competition with all plants will be selected. In the managed agricultural biosphere, on the other hand, growers and plant breeders will select the plants whose characteristics enable growers to obtain the most profit in the future high-CO_2 world. Those plants and crops will be grown. Therefore, the plants that possess the largest yield response to CO_2 will likely be selected, and such a yield can be regarded as an adaptation to a higher-CO_2 world. This gives us an indication that on a long-term basis, both in natural condition and in agriculture, there is likely to be change in composition of green biomass. Other than CO_2, there could be a change in green biomass composition from other climate change-induced variation of environment. Let us discuss a few of those variations.

7.3 Climate Change-Induced Variation

Although all species may respond to increase in CO_2 concentration, there would be considerable interspecific variation as mentioned before. The prominence of variation will depend on geographical area and the nature of change experienced by that area. Predictions from climate models indicate that there may be a general warming with the poles heating more than the equator. Considering that production worldwide is limited more by low than by warm temperature, climate change-related warming together with longer growing seasons should enhance the production. But as the temperature regimes of the production systems will shift toward the pole, the soil type may not be changed within the time frame; hence, the new crops under that temperature regime will face different soil types and will have to compete with production crops that existed before.

Along with a warming earth, the climate model generally also predicted greater evaporation from tropical oceans in the future high-CO_2 world, and, consequently, average precipitation is also predicted to increase worldwide. However, the water content of an air mass depends on where it has been as well as where it is. Therefore, the greatest concern is focused on whether a higher-CO_2 world, with possible higher temperature, will affect the pattern of atmospheric circulations and consequent pattern of temperature and precipitation distribution. Manabe and Wetherald (1980), for example, used a general circulation model and predicted that some areas of the earth might indeed become drier while others become wetter. Kimball (1985) stated that an increased atmospheric CO_2 concentration may ameliorate water and salinity stress through climate change, so, marginal semiarid land may even become more productive. However, a study of the pattern of present-day precipitation suggests that the prediction of the model is not reliable. The future weather patterns have not been predicted with any certainty. Some authors like Idso (1984) did not find any significant changes in temperature increase as it was predicted by GCM. However, abrupt changes in temperature and precipitation are happening in different areas because of the increasing atmospheric CO_2 concentration.

Nevertheless, if changes in weather patterns do occur, then the shift of weather will affect plants indirectly. Keeping aside any of the direct effects of CO_2 on plants, these shifts would be strong enough to shift the production areas. Even if agriculture can adapt to climate changes, it is probable that such changes may limit production to levels below the average potential yield increase of about 32% that appears possible with a doubling of CO_2 concentration (Kimball 1985). So, in the long run due to climate change, there would be a shifting of vegetation composition; however, such shifting may not happen with necessary increase in production for a setback on emission. Moreover, such a change in composition of green biomass could bring a socioeconomic change that could agitate human beings' contribution to emission. Considering human beings as the center of emission problem, the socioeconomic shifting merits some discussion here.

7.4 Aspects of Socioeconomic Conditions

We have come to know that the changes in vegetation responses and capitalizing them as a sink of CO_2 cannot bring a sustainable solution. The most important strategy for resolving the greenhouse problem is to reduce or stop it. This requires the integration of scientific issues and political and socioeconomic factors. An increase in population and the corresponding demand for food and energy may have an induced impact on climate change. The technology is developing to meet the increased demand for food and energy. However, at the same time, energy use as well as CO_2 emission into the atmosphere is increasing proportionately. Though C filter techniques have been developed to reduce the industrial carbon emission, it is not helping much because disposal of waste C appears as another problem. Tree planting may have a minor role to play in controlling the effect.

Marland (1988) has calculated that it would require 700 million ha of forest growing rapidly in a warm-wet environment to absorb worldwide excessive fossil fuel-related CO_2 emissions. This is an area roughly the size of Australia, equivalent to the area deforested in the whole of human history. Moreover, when a forest reaches maturity, it must be harvested, stored permanently without decomposition, and new plantation commenced to achieve the benefits. So the main solution to the greenhouse effect remains in the reduction of fossil fuel use. New technologies might need to be developed to supplement the supply of energy. But such a development has inbuilt risks of high cost and achievability. As the problem is global, isolated actions may bring little consequence in reducing the emission of greenhouse gases albeit a long-term solution cannot be expected.

One of the goals of the Toronto conference was to reduce the CO_2 emission by 20% by 2005, and subsequent conferences prescribed to come back to 1990 level. They have not been achieved. The greatest potential as suggested by Kolm (1990) lies in energy management where cost economies, fuel conservation by advanced technologies, and some manufacturing opportunities focus on:

1. Compact motor transport
2. Advanced combustion technologies
3. Improved energy efficiency (energy cascading) in industry
4. Low-energy electrical appliances
5. Energy efficiency in housing and urban planning
6. Alternative energy, bioenergy, and other renewable energies

Presently the developed countries are working toward them. Potential impetus has been given to alternative and renewable energy. However, the goal for reducing emissions cannot be achieved if transferring the preventive technologies from developed to developing counties is not quick enough. These technologies could be based on anything or a combination of them helping the situation of CO_2 emission. Cost-effective technology for using biomass fuel and renewable energy needs to be developed to reduce the use of fossil fuel or fuel wood. At the same time, considerable management effort has to be given to reduce the conflicting issues of energy production with food security situation.

7.5   Aspects of Management Change

As we have said earlier, human being cannot change the physical environment directly. Their effort is limited to managing biotic components to cause changes in physical environment. By now, we understand that a biotic component takes a long time to have a considerable effect on physical environment. Thereby through management issues, it is unlikely that short-term resolution might be achieved unless a drastic cutoff in emission happens. A long-term resolution could be targeted through appropriate management of green biomass and required changes in technology. Such a technological change would be possible depending on whether more attention is given to future forecast of climate change impacts rather than the socioeconomic priority. Some of the aspects are mentioned below depending on the scenario of available information.

7.5.1   Plant Selection and Breeding

Humans usually select plants for higher yield and other desirable traits. Within the last couple of centuries, plant breeders have accelerated the development of desired plants. While annual species may have been acclimatized readily, new breeds have been developed through genetic modification technologies. They have not been targeted to reduction of emission but to production for meeting the increased demand of human being. Not much has been done toward trees that can keep C locked for a longer time. We have already shown that there are variations in adaptability between the C_3 and C_4 plant groups. But modern genetic shearing techniques have made it possible to change C_4 plants to C_3 plants, and in the near future, it is reasonable to expect that technologies, such as genetic engineering and tissue culture, will make it possible to achieve varieties of species capable of utilizing the potential for increased biomass growth at increased CO_2 concentrations. However, there is no guarantee that such discoveries will reduce N input and other nutrients significantly. As a consequence, N-driven emission problems are likely to drag the climate issue backward.

7.5.2   Cultural Practice

Cultural practices include fertilizing, applying pesticides, trimming and pruning, and irrigation for increase of productivity. These practices may need to be attended along with the CO_2-induced change of vegetation for minimizing the backdrop effect. Some of these aspects are described below:

7.5.2.1   Fertilizer

From C/N ratio discussion, we understand that for obtaining the maximum benefit from higher atmospheric concentration of CO_2, soil nutrient levels must not be depleted. Fertilization will probably become more important than the genetic modification. But it is not clear what ratio or extent of increase of fertilizer use would be necessary. From some reports, it could be estimated that use of adequate fertilizer would reduce the period of attaining the final biomass, at least in some cases, which means the crop rotation would be reduced, but the final biomass would remain the same. As a result, a larger number of crops could be possible in a year through managing carbon nutrient relationship. Hence, to attain the maximum biomass, fertilizer management would be necessary along with genetic modification. Some elements of fertilizer could be more essential than others depending on the natural fertility of the soil and the crop requirement. However, long-term consequences of such an induced fertilizer management to soil quality are not very clear.

7.5.2.2   Pest Control

Pests to the economic crop may be plants such as weeds, insects, and microorganisms. The nature of the infestations could be changed due to CO_2-induced effects and adaptability. We now hear that millions of hectares of pine forest of British Columbia (BC) have been damaged by pine bark beetles which were never before heard of in that area. It has been expected that bark beetle has migrated there and established a new habitat due to climate change related to warming. It will take centuries to get these BC pines to get adapted against those beetles to help themselves become established again in the succession. Similar changes were anticipated in the attack of Asiatic long-horned beetles on Ontario ash forest. Therefore, whatever changes we expect from genetic and nutrient management, in the long run, they might not be useful if climate-related changes are not attended to adequately.

7.5.2.3 Weeds

The pattern of weed competition is expected to change due to an increase in the concentration of CO_2 for the following reasons:

1. Effect of CO_2 on growth pattern
2. The geographical shifting of ecological regimes of the species
3. Changes in the root exudates chemistry and quantity which may stimulate the associated organisms to utilize more nutrients

Because weeds are mostly C_4 plants, and are ephemeral in nature, they involve a larger number of rotations in a given period of time, and they have more diversities than the crops. Therefore, they are likely to be quickly adaptive. Ziska et al. (2004) has reported some changes in root-shoot growth ratio of *Cirsium arvense*. In addition, due to climate change-related increase in temperature, the weeds are likely to shift toward higher latitude and likely to become invasive in their new habitat. Although chemicals presently used are adequate for controlling weeds, under changed circumstances, they will no longer be adequate and will lose their specificity. Applying excess amounts of chemicals for controlling the weed will not be useful for the environment at all. As a result, a change in cultural practices would be necessary. More attention can be given to integrating biological control; however, the researches in that area are not adequate. This is because we see such researches as scientific not a climatic one.

7.5.2.4 Insects and Diseases

A doubling of CO_2 concentration is not likely to have any direct effects on insect or disease organisms. Because CO_2 will affect their host plants, some interactive effects like feeding and breeding habits will occur undoubtedly. It is expected that insects and disease will simultaneously evolve and adapt to the future composition of green biomass. For example, if more nectar is produced in the flower due to an increase in CO_2 concentration in the atmosphere, then bees and other pollinators will probably respond with increased populations and pollinating activity. It is not clear still whether plants grown with CO_2 enrichment are likely to be more or less vulnerable to insect attacks. Lincoln et al. (1984) reported from an experiment with soybean grown under different concentration of CO_2 and with soybean looper larvae that the larvae fed more on the leaves grown under higher CO_2 concentration. On the other hand, correlation between disease resistance capacity and increase in CO_2 concentration has not been studied at all. Due to changes in the seasonal pattern, the trees may shed their leaves early due to which larvae of many insect will starve and will not grow to maturity. Including all these changes, as with insects, it is logical to expect that mutual host-pest adaptation will occur. In this case, some diseases will become more prevalent than others. But as the associated climate change may affect the pest significantly, it is too early to talk about the host-pest relationship at elevated atmospheric CO_2 concentrations.

7.5.2.5 Water Use

We have mentioned before that an important different direct effect of increased CO_2 concentration is the tendency to reduce the stomatal conductance and hence the rate of transpirational water loss from the surface of the leaves. Kimball and Idso (1983) calculated 34% reduction in transpiration with doubling of CO_2 concentration from the results of 46 observations from 18 different species in short-term growth chamber experiment. Such a reduction in transpiration represents an upper limit to the amount that evapotranspiration may be reduced in future vegetation by a doubling of atmospheric CO_2 transpiration. However, plants are probably going to be larger and have a greater leaf area in the future high-CO_2 concentration world, thus tending to increase transpiration. They probably will also have larger, more vigorous root systems and be able to extract more water from the soil. Any decrease in transpiration will make more thermal energy available for soil evaporation. Therefore, the amount by which the consumptive water use will be reduced by a doubling of CO_2 concentration is going to be less than the potential average 34% reduction in leaf transpiration rate (Kimball 1985).

Rogers et al. (1984) measured the water use for a few days from well-watered, mature soybeans grown in pots in open-top chambers at several CO_2 concentrations. They found that a doubling of the CO_2 level caused a 40% reduction in the rate of water loss expressed as either as per leaf area or per pot. In this study, there was a 67% reduction in stomatal conductance and a 26% increase in leaf area. So, in that experiment, the change in conductance dominated the change in leaf area. Morison and Gifford (1984) observed from a study in a growth chamber with 16 different species grown in pots that the rates of water loss per pot were not significantly different between ambient and doubled CO_2-concentration treatments. Leaf areas were increased to 20–75%. So, apparently, decrease in stomatal conductance and increases in leaf area and leaf temperature were balancing factors as far as leaf transpiration is concerned. Findings of Kimball (1985) also support the idea.

From the studies, it could be expected that consumptive use of water will not be greatly affected by the future high CO_2 concentration, although a little reduction of water use could be expected. The rate of reduction will probably depend on the increase in the leaf area. As stomata of C_4 plants are responsive (described before) but photosynthesis relatively remains unaffected, C_4 plants may have a greater advantage than C_3 plants in terms of water use. As a result, dominance and diversities of C_4 plants are more expected than those of C_3 plants in the climate-change world. We also understand that more than one environmental phenomenon would be associated with these responses of plants. Let us explain these situations of environmental regulations in a different chapter.

References

Callaway JM, Currie JW (1985) Water resources systems and changes in climate and vegetation. In: White MR (ed) Characterization of information requirements for studies of CO_2 effects: water resources, agriculture, fisheries, forests and human health. DOE/ER-0236. United States Department of Energy, Washington, DC, pp 24–67

Enting IG, Mansbridge JV (1989) Seasonal sources and sinks of atmospheric CO_2: direct inversion of filtered data. Tellus 41B:111–126

Idso SB (1984) The case for carbon dioxide. J Environ Sci 27:19–22

Kimball BA, Idso SB (1983) Increasing atmospheric CO_2: effects on crop yield, water use, and climate. Agric Water Manag 7:55–72

Kimball, BA (1985) Adaptation of vegetation and management practices to a higher carbon dioxide world. In: Strain BR, Cure JD (eds) Direct effects of increasing carbon dioxide on vegetation. United States Department of Energy, Washington, DC, pp 185–204

Kolm JE (1990) The potential for reduction of carbon-di-oxide emissions in Australia. In: Swaine DJ (ed) Greenhouse and energy. CSIRO, Melbourne, pp 69–80

Lincoln DE, Sionit N, Strain BR (1984) Growth and feeding response of *Pseudoplusia includens* (Lepidoptera: Noctuidae) to host plants grown in controlled CO_2 atmospheres. Environ Entomol 13:1527–1530

Manabe S, Wetherland RT (1980) On the distribution of climate change resulting from an increase in CO_2 content of the atmosphere. J Atmos Sci 37:99–118

Marland G (1988) The prospect of solving the CO_2 problem through global reforestation. US Department of Energy, DOE/NBB-0082, US Department of Commerce, Springfield, Virginia

Morison JIL, Gifford RM (1984) Plant growth and water use with limited water supply in high CO_2 concentrations – leaf area, water use and transpiration. Aust J Plant Physiol 11:361–374

Rogers HH, Sionit N, Cure JD, Smith JM, Bingham GE (1984) Influence of elevated carbon dioxide on water relations of soybeans. Plant Physiol 74:233–238

Tans PP, Fung IY, Takahashi T (1990) Observational constraints on the global atmospheric CO_2 budget. Science 247(4949):1431–1438

Tucker GB (1990) Scientific uncertainties associated with the greenhouse problem. In: Swaine DJ (ed) Greenhouse and energy. CSIRO, Melbourne, pp 11–23

Ziska LH, Faulkner SS, Lydon J (2004) Changes in biomass and root-shoot ratio of field grown Canada thistle (*Cirsium arvense*), a noxious, invasive weed, with elevated CO_2: implications for control with glyphosate. Weed Sci 52:584–588

Chapter 8
Environmental Regulations on Activities Associated with Enhanced CO_2 Concentrations

An attempt has been made in previous chapters to describe the impact of an increase in atmospheric CO_2 concentration on vegetation growth. The climate change effect is expected to bring some complex secondary changes in a series of environmental factors, and these changes may play a significant role in biomass production as well as in maintaining genetic diversity. Climatic change would influence biodiversity in two main ways:

- First, through changes in habitats associated with changes in land use and in fire regimes
- Second, through direct effect on the species concerned, that is, the ecological range of different species may change

For example, a 2.5°C increase in temperature is expected to greatly increase the distribution and abundance of *Eucalyptus grandis* and almost eliminate *Eucalyptus radiata* from the north coast of New South Wales of Australia (Walker et al. 1990). Though the degree and direction of change in climatic factors on biomass growth under different conditions are not certain, the general impact of possible changes in those factors of biomass growth under different conditions is discussed. The impact on biomass growth may be quite complex to describe when changes in all the factors occur simultaneously. Several review articles report the occurrence of greater tissue growth during periods where photosynthesis rates were small or nil, which means that the impact of CO_2 on biomass growth is delayed action, but the opposite has also been reported.

CO_2 buildup in the atmosphere may have an impact on vegetation growth through one of the following changes in climate influencing the water cycle:

- Effects of change in climatic variability such as temperature, precipitation, evaporation, and other metrological variables
- Effects of changes in the frequency of occurrence of extreme events such as floods and droughts
- Effects of change in the distribution of events over time
- Effects of different rates of climate change

M. Ali, *Climate Change Impacts on Plant Biomass Growth*,
DOI 10.1007/978-94-007-5370-9_8,
© Springer Science+Business Media Dordrecht 2013

The changes in these variables may have positive or negative effects depending on other environmental conditions and the nature of vegetation. Some simulation studies of individual plant impacts on community models have been done by Aston (1984) and Idso and Brazel (1984). The findings show that these variables may interact with each other and may have a feedback effect on vegetation growth. The combined impact of these variables at the plant community level is still not clear. We will try to describe some of them here in an isolated way.

8.1   Temperature

Temperature change, one of the prime environmental factors, is expected to bring complex changes in vegetation. Usually, moderate daytime temperatures favor shoot growth, whereas lower night temperatures restrict growth. Luxmoore (1991) has observed shoot growth rate to lag 2–5 h behind the air temperature changes in the diurnal cycle. There is a lag of several hours in the diurnal maximum soil temperature, and this may favor the root growth during afternoon and evening. This growth activity demands energy from the plant store whereas photosynthesis is conducted in moderate daylight to fill up the store. At high temperature, stomata usually close to save moisture loss, and hence, it could be expected that the best CO$_2$ utilization takes place at moderate temperatures. The changes in temperature are expected to bring changes in the diurnal cycle as well as seasonal cycle resulting in changes of above-ground biomass to below-ground biomass ratio. These changes could end up in the change of dominance of some groups of plants over others. The end result could be in possible changes of species composition and the geographical shift of species.

8.2   Precipitation/Water Availability

Wetting and drying cycles may have a marked effect on the CO$_2$ utilization by the plant. This variation of the cycle depends on climate. When soil is dry, root growth is impeded, and a chemical signal is released from roots and transported to foliage where it induces stomatal closure. This mechanism provides feedback from the root sink that reduces the photosynthesis and transportation by impeding gas exchange. Water stress is also a dominant limitation to leaf growth of forest trees (Gholz et al. 1990) which may hamper the photosynthetic activities. Generally, cell syntheses (cell growth, wall synthesis, protein growth, etc.) are very much dependent on water availability. On the other hand, these are the activities which use the energy produced from CO$_2$ and contribute to biomass growth. Hence, additional CO$_2$ may have little impact on biomass growth in a drought condition.

Diurnal variation in leaf water potential between a maximum of about –0.2 and a minimum of about –2.0 MPa is typical for many tree species (Klepper 1968). Thus,

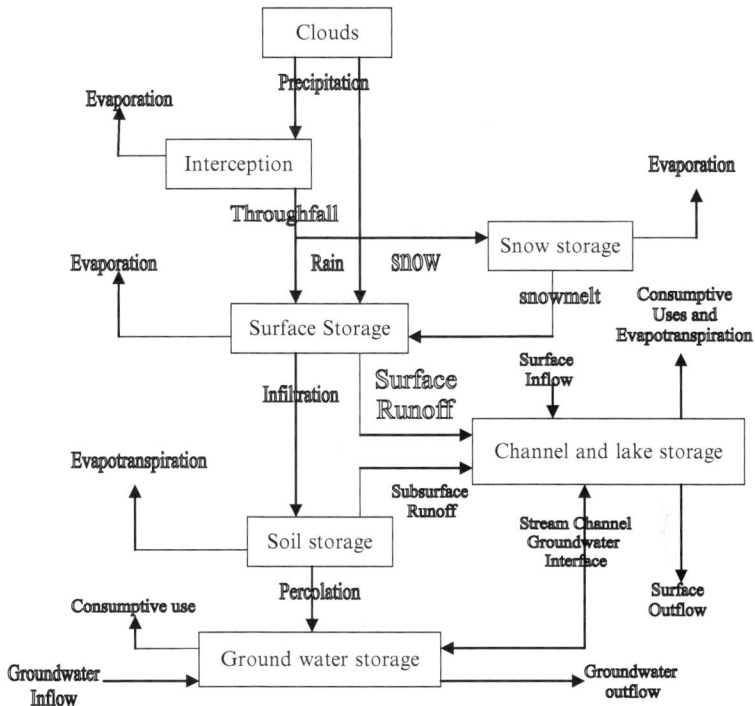

Fig. 8.1 Schematic diagram of the hydrologic system of a drainage basin (Source: Callaway and Currie (1985))

during the daytime, shoot water potential usually permits photosynthesis, although for much of the time, it may be too low for cell growth. This is supported by the findings of Cremer (1976) that diurnal pattern of the shoot elongation of *Pinus radiata* and *Eucalyptus regnans*, with growth occurring mainly at night following hot sunny days. However, the elongation growth may occur mainly in daytime in the mild and wet days as well. Conditions that favor growth during the day may result in enhanced photosynthesis because feedback inhibition is reduced. For example, when it is overcast or raining, growth and hence photosynthesis may be greater than a hot sunny day when tissue water potential is low. These changes could result from the changes in the climate or the hydrologic cycle.

We know that growth of vegetation is dependent on the hydrologic cycle. Many scientists have predicted that the hydrologic cycle will vary due to climate change effect. The variation may be different from region to region and may have different effects depending on the other environmental conditions, and the effect may be positive or negative. Hence, increased concentration of CO_2 may affect vegetation growth through the hydrologic cycle differently in different region. Let us take the hydrologic cycle as suggested by Callaway and Currie (1985) to explain the state and nature of effects on vegetation growth (Fig. 8.1):

It is thought that due to an increase in temperature, the moisture-holding capacity of the atmosphere may increase, and, as a result, rainfall may increase in certain conditions. If there is a lot of vegetation on the earth's surface, most water will reach the earth's surface as through fall after interception of precipitation by the vegetation canopy. If there is little vegetation, most precipitation will come directly to the surface. The question is not whether the water will reach the surface but how much water will be infiltrated into the soil. If the precipitation comes directly to the soil, most of the water will flow directly to the lake or sea as surface runoff and will also cause massive erosion and further degradation of the conditions of vegetation. On the other hand, if there is dense canopy through fall, water will infiltrate into the soil and may be stored for later use for vegetation (soil storage). Subsequently, water from soil storage may return to the atmosphere through evapotranspiration which may contribute effectively to cloud formation and the processes of the hydrologic cycle. The excessive water will percolate slowly to the ground storage.

From the description, we see that the soil storage and underground storage of water will largely depend on the density of vegetation. However, the stimulation of vegetation growth by CO$_2$ concentration in the atmosphere will also partially depend on the extent of existing vegetation on the earth's surface. On the other hand, if there is a lot of vegetation on the earth, the intensity of greenhouse effect may be restricted. So the interaction is very complicated to be described by a single model. If the rainfall is reduced due to greenhouse effect, then perhaps the effect on vegetation growth will be deteriorating, cumulative, and ultimately will be irreversible, that is, a negative effect. The complication will increase further if the duration and distribution of precipitation are included in the explanation.

8.3 Effects of Elevation and Aspect

As elevation increases, the effect of CO$_2$ on vegetation may change because, at a higher elevation, the density of vegetation canopy decreases and the importance of transpiration also declined. At higher elevation, the effect of CO$_2$-induced change on evaporation will probably be stronger than the fertilizing effect of CO$_2$ on transpiration (Callaway and Currie 1985). Hence, the contribution to biomass growth could be less at higher elevations than at lower elevations. Other environmental factors, such as nutrient and moisture availability and soil physical properties, may be too limiting to facilitate much CO$_2$-induced growth at higher elevations. These properties depend on the degree and direction of slope. For example, elevated areas usually experience longer diurnal cycles, and more rainfall, but on the slope away from the sun and on leeward slopes, there may be an opposite effect. Hence, the impact of CO$_2$-induced growth will vary from slope to slope depending on their aspect and degree; however, where there are lots of complexities of climate change to consider, considering the impacts of slope on overall changes of vegetation may not be very significant at global scale.

8.4   Effects of Drought Conditions

The drought condition, if it occurs, may have a very sensitive effect on vegetation growth. During drought, the volume of moisture withdrawn from storage by transpiration is likely to increase, and the soil moisture level may decline. As soil moisture becomes depleted, several things happen:

– Vegetation begins to wilt.
– Density of vegetation canopy decreases.
– Soil becomes more exposed to direct sunlight.
– More water would be required to grow under this condition.

Eventually, some plants in the vegetation will die, and if drought conditions are prolonged, changes in soil structure and chemistry will begin to take place with the onset of desertification. Under that situation, the potential of future vegetation growth may be reduced significantly resulting in a permanent desertification which may cause an irreversible effect.

Under drought conditions, the surface temperature of the soil will become high resulting in a higher evaporation from the soil surface, and the top soil may become loose and detached. Under that condition, rainfall, if it occurs, can increase the sheet erosion, and this in turn may affect the health and composition of the vegetation. Under some other conditions, when the evaporation is high, the water from the aquifer may rise depositing salt within the soil. If there is little precipitation, the soil could become saline and subsequently affect vegetation growth. In general, salinity has a negative correlation with the precipitation and a positive correlation with the temperature (Callaway and Currie 1985). Therefore, under drought condition, the soil will unveil different reactions that could be unacquainted with existing vegetation and could result in a dramatic change to desertification.

8.5   Effects of Canopy Cover

By now, we understand that the vegetation canopy plays a central role in regulating the principal moisture and energy fluxes between the atmosphere and the land surface. It also modifies the soil structure through the addition of litter and by protecting the soil from erosion and leaching. Under laboratory conditions, the experimental result is that an increase in CO_2 results in more vigorous plant growth, increased dry-matter production including fruit and leaf area, greater water-use efficiencies of plants, and changes in competition between plants (Solmon and West 1985). Some of the effects are discussed in the following paragraph considering the forest as the representative of canopy.

8.6   Effects of Forest

Forests are normally so densely populated that usually trees have to compete with each other for sunlight, water, and nutrients. Also, tree species have different resources and use strategies in order to grow and reproduce in the presence of other trees. Thus, most trees grow in the presence of stress, making them sensitive to environmental changes of the kinds expected from CO_2-induced climate change.

Completion of the life cycle of a normal tree can be viewed as a three-part sequence:

- First, seeds germinate, take root, and survive as seedlings.
- Second, the seedling reaches the uppermost level of forest where sunlight is freely available.
- Third, the trees survive long enough to reproduce themselves.

The environmental variables, for which the trees compete, for example, nutrients, water, light, temperature, will differ in importance at different stages on the life cycle and in between species. As a result, the point within the life cycles at which CO_2-induced climate changes may affect tree growth will vary from one species to another as well as from one place to another within the geographical range of each of the species. If forest ecosystems respond to CO_2 fertilization, they will do so as a function of changing competitive advantage among species. Though research data shows that increase in CO_2 concentration could benefit trees, information on the response of long-lived trees in forest to the greenhouse effect is lacking. More fixed carbon may be produced, but over the years, there may not be enough nitrogen and other nutrients to meet the new demand. Photorespiration may decrease, but it may not decrease enough to offset increase in dark reaction in response to soil and air temperature increases.

Water-use efficiency may increase in individual leaves, but the increased load of leaves generates different total evapotranspiration per plant. Photosynthesis may be more efficient in the shade, but it may not compensate for the increased shading because of more and thicker leaves. Indeed, plants acclimatize for several days to increased CO_2 and then may cease to respond to the induced growth (Solmon and West 1985). Hence, there is a doubt about the long-term implications of short-term experimental results. In fact, Solmon and West (1985) reported in a review that the growth of trees had been reduced where the trees were affected by acid rain or gaseous air pollutant though the CO_2 concentration had increased about 25–30% since 1850. Therefore, the combined effect of climate change-induced environmental regulation may not likely bring added benefit from CO_2 increase.

8.7   Ecological Issues

Ecological issues are inseparable from those related with the carbon cycle, climate, and vegetation responses to CO_2 increase. For example, reproductive success will determine what trees will survive and where. Seed source variability, along with the soil moisture and temperature variance, controls tree seedling establishment. But the extents of change in moisture and temperature as well as magnitude of the effects are not yet clear. Growth and maturity of established seedlings and the geographic range of tree species depend on the absolute amount, distribution, and variance of precipitation. Under climate change situations and on the extremes of temperature – none of these has been predicted accurately.

Availability and requirements of nutrients for tree growth depend on population and activity rates of decomposer organisms among other things (e.g., quality of litter, type of decomposer community, species composition). Decomposer population and their activity rates, in turn, depend on the status of soil temperature and moisture. Specifications of these variables and their interaction are difficult. The opposite of growth is senescence and tree mortality. Senescence occurs when tree vigor declines, growth of root system slows, and the water and nutrient requirement of leaves cannot be met adequately (Kramer and Kozlowski 1979). If senescence behavior changes, there is likely to be a change in microbial responses resulting the change of soil condition and pathogenic activity.

Chronic mortality is similar to senescence mortality, although it occurs when the growth of any tree is slowed in successive or near-successive years. At this point, the tree can no longer resist normal stresses. Premature senescence and increased mortality rates may be reduced if rising atmospheric CO_2 levels enhance annual growth. The difference between early and late senescence could affect the carbon source and sink properties of the terrestrial biosphere as well.

Catastrophic mortality is more visible than chronic mortality. Forest losses from, for example, wildfires, windstorms, and floods, are directly related to local and regional geographic distributions of climate extremes. The impact of atmospheric CO_2 increases on these climatic extremes is very difficult to project. Pathogen and insect epidemic causing local or regional extinction of whole tree plantations are thought to be climatically induced. Future CO_2-induced changes may create these kinds of events resulting in an ecosystem change.

The rates and outcomes of forest development (succession) are generally understandable under constraints imposed by current atmospheric CO_2 concentrations, climate variance, species composition from seed sources, and disturbance frequencies, intensities, and aerial extents, all of which vary with local and regional geography. However, whether succession rates will slow and local dieback will occur, if climate changes are continuous, is uncertain. It is also not known whether CO_2 will enhance the rate of succession by relieving growth-loss stress or by the future local and regional distribution of climate variance and forest disturbance by wind, fire, and flood known. It is also uncertain what species will be available to influence forest succession in the future. A set of known and unknown issues due to CO_2-induced changes are present in Table 8.1.

The descriptions in Table 8.1 demonstrate that the indirect effects of climate change are likely to be more apparent in forest behavior than will the direct effects of carbon fertilization. The table also reveals that the vegetation response to CO_2 concentration is a function of the ecosystem. Some of these cited in the table are findings of simulation results of assume doubled CO_2. Solomon and West (1985)

Table 8.1 Primary knowns, uncertainties, and unknowns requiring resolution in order to appraise forest ecosystem responses to climate change induced by CO_2

Issues	Knowns	Uncertainties	Unknowns
1. Carbon cycle issues			
(a) Atmospheric CO_2 accumulation	Atmospheric CO_2 concentration is rising	Preindustrial CO_2 value	Time when doubled and maximum CO_2 will occur
	Seasonal CO_2 amplitude is increasing	Source of increasing amplitude	Maximum future CO_2 concentration
	Land use (forest clearance) contributes to atmospheric CO_2	Rates at which land use generates atmospheric CO_2 increases	Significance of amplitude changes to carbon cycle, vegetation, and climate effects
			Future contribution to CO_2 by land use
(b) Atmospheric CO_2 depletion	Oceans are the primary global atmospheric CO_2 sink	Rates at which CO_2 crosses the ocean-atmospheric interface by geographic region	Internal circulations of oceans by geography that control up- and downwelling CO_2-rich and CO_2-poor water masses
	Terrestrial biota is a secondary global atmospheric CO_2 sink	Current areas and amount of C stored in terrestrial biosphere	Future C source, sink, and storage properties of terrestrial biosphere
2. Climate effect issues			
(a) Geography of climate change	Greater warming at poles than at equator	Intensity of latitudinal temperature differences	Local and regional temperature shifts
	Change will occur in geography of precipitation	Geography of precipitation	Local and regional precipitation shift
(b) Variance from mean climate changes	Variance may or may not change	Presence or absence of variance changes from current geography	Amount of variance change, if any, locally, regionally, and globally
(c) Temporal chronology	Climate will change over time	Rate and nature (step function, climate continuous linear, etc.) of climate change	Temporal chronology of climate change
(d) Seasonal features of climate change	Greater warming in winter than summer	Amount of seasonal warming	Local and regional change in seasonal warming. Local and regional time difference between summer and winter warming
	Quicker warming in summer than winter	Time difference between summer and winter rates of warming	

3. Vegetation effect issues			
(a) Application of single-factor greenhouse experiment to multifactor field situation	Published report on single-factor experiment shows increased photosynthesis, decreased water use, increased leaf area, increased dry weight, and increased photosynthesis in shade	No published reports on multiple-factor experiments that realistically simulate field conditions; whether nature trees and ecosystem will respond similarly	Amount, if any, of response in multifactor situations; quantitative importance of offsetting (canceling) factors to tree growth
(b) Application of short-duration (hour, day) fumigations to long-term growth	Published reports on longer-term (months, seasons) experiments show short-duration effects, decline, and/or disappear	Will field-grown trees cease responding to CO_2 increase?	How much, if at all, will field-grown trees acclimatize to enhanced CO_2?
(c) Applicability of experiments on herbaceous annuals, leaves, branches, and tree seedlings to mature trees	Effects described in 3a are found in these plants or plant parts	Will mature trees respond as do plant parts tested? If so, what is the nature of forcing function (linear, logarithmic, etc.)?	Amount, if any, by tree species or by other functional unit (shade tolerance, growth rate, etc.) that trees will respond as do other plants or plant parts
(d) Applicability of experiments in 3a, 3b, and 3c to regional or global vegetation	The few trees and fewer species examined to date have declined in growth, while CO_2 has increased in concentration	Has the current 30% CO_2 increase affected tree growth; importance of climate change and other pollutants during same period in depressing tree growth	Will future changes in climate and atmospheric pollutants negatively affect tree growth? If so, how much compared with possible CO effects?
4. Seedling survival	Temperature, precipitation, variance control, and initial seedling survival	Will temperature and precipitation variance change in the future?	By how much, if any, will variance in temperature and precipitation change?
	CO_2 enhancement, below current forest floor concentrations, increases vigor, production, biomass leaf area, and drought tolerance differently by species.	Will CO_2 enhancement above current forest floor concentrations have the same effects?	Will CO_2 increase have long-term effects on massive mortality, and if so, which species will be favored?
	Pathogens and predators destroy seeds and seedlings	Will pathogen and predator effects be different in the future?	Which species will be favored and by how much if predator and pathogen effects occur?

(continued)

Table 8.1 (continued)

Issues	Knowns	Uncertainties	Unknowns
5. Sapling and tree growth	Climate variance controls geographic ranges of mature species	Specific climate variables and their variance values controlling the cold, warm, and dry parts of each species' geographic range	Future local and regional climate variances, future effects of CO_2 on responses of species to climate variance, and future geographic range of species
	Climate variance controls annual growth rates	Specific climate variables and their variance values	Future local and regional climate variances; resultant shifts in competitive advantages among species
	Nutrients for growths depend on decomposition rates	Will decomposer populations change nutrients available because of warming of soils?	Future amount of nutrient available because of changes in decomposer activities due to warming
	Root respiration depends on soil temperature, water, and oxygen availability	Will root respiration increase because of climate change?	If root respiration changes, will it counteract increased production from CO_2 if such increase occur?
6. Death and reproduction			
(a) Senescence	Senescent trees die from stresses that younger trees can survive	If climate changes continuously, will stress become prematurely senescent because of the additional stress of this continuing change?	Will CO_2 induce vigor and prolonged tree life, and if so, which species will be most affected and which will be least affected?
(b) Catastrophic age-independent mortality	Wildfire frequency and intensity depend on climate, fuel and time since last fire	Geographic changes in wildfire frequency and intensity due to future climate change	Local regional changes in the temperature and precipitation extremes that control fire frequency and intensity
	Windstorm and flood storm depend on frequency of rate climate extremes, topography, and land use	Geographic change in land use due to climate change	Local regional distribution of future and precipitation wind extremes
	Pathogen and insect epidemics occur rarely	Climate dependency of pathogen and insect epidemics	Will climate control pathogen and insect epidemics, and, if so, how?

(c) Chronic age-independent mortality	Occurs when production is too slow to provide enough fixed carbon for metabolism, slowing growth and reducing resistance to stress	Qualification of "too slow" by species of tree	Will CO_2 experiments with seedlings apply to mature trees, reducing chronic age-independent mortality; future local and regional distribution, frequency of climate variance, and slow growth
	Chronic insect predation on trees is a normal stress that vigorous trees survive	Will chronic insect attacks change because of changing climate variance and changing forest composition?	Climate variance that controls insect population sizes and species composition; forest composition that controls the same
	Air pollutant damage (gaseous, acidic precipitation) is an increasing stress from which vigorous trees previously survived	Future air pollutant levels and loss of tree vigor by species to future air pollutant levels	Future combined effect of unknown air pollutant levels; climate extremes, insects, pathogen prediction, and CO_2 effects on forest growth
(d) Plant succession	Rates, outcomes of succession under current constraints by climate variance, seed sources, disturbance frequencies, and CO_2 concentrations	Will rates slow because of stress of possible continuous climate change; presence, intensity of forest dieback, and recovery	Will CO_2 enhance rates by relieving stress; future local and regional distribution of climate constraints (variance); future disturbance by wind, fire, flood
(e) Plant migration	Trees migrated 100–400 m/year in the past 10,000–15,000 years	Future migration rates on a land use dissected landscape; survival of seedlings of species ill adapted to new conditions	Availability of seed sources with time, availability of empty niches, and tree planting

Source: Adapted from Solmon and West (1985)

from their informal researches have observed that in some ecosystems, the net result is carbon sink (closed forest), and in some others, it appeared as source of CO_2 (Prairie forest in North America) with some percentage (5%) degradation of standing biomass.

There are also questions concerning the effective length of the growing period. It is reasonable that a warmer climate will lead to increase in the growing period within lethal limits. Interactions exist between the direct and indirect effects of increased CO_2 concentrations in agriculture, which involve the rates of photosynthesis and respiration, water-use efficiency, and nutrient transport within the plants. Plants respond differently to increases of ambient levels of CO_2 depending on their photosynthetic metabolic pathways.

Plants are classified as C_3 (rice, wheat, barley, oats, soybean, sunflower, and alfalfa) and C_4 (corn, sugarcane, sorghum, and millet) and CAM type (pineapple and cacti). Most food crops belong to either C_3 or C_4 plants. The increase in CO_2 concentration generally should enhance photosynthesis, with the increase being more marked for C_3 plants, and the increase photosynthesis probably will increase the demand on nutrients and fertilizer. On the other hand, transpiration rate is expected to be reduced by both types of plants and will be particularly pronounced in C_4 plants due to enhanced levels of CO_2 (Decker et al. 1985). These results are from laboratory studies. In the field condition, the effects on agricultural crops are likely to be the same. The region of crop production may migrate with the change of climate.

Temperature, humidity, atmospheric circulation, and other weather parameters influence pests by affecting their development, distribution, migration, and adaptation. For example, *Phytophthora* spp., causing root rot in Eucalyptus or causing potato blight, develops under cool moist conditions (Agrois 1978), but *Macrophomina phaseolina* (charcoal rot), another fungal pathogen, could thrive in hot, dry condition. Pest development depends on the temperature. The geographic distribution of pests is more or less controlled by the climate and the climate changes, so, the distribution of pests will tend to change under CO_2-induced climate.

Some pests migrate along the same routes each year (e.g., locusts); others appear to be transported by low-level atmospheric circulation, for example, cutworm moth. The reproductive strategy of pests also depends on the environment. Insect population generally increases with the increase of temperature, but some insects become inactive at higher temperature and some do better in dry conditions. In general, the change in the pest and insect population may become complicated. Though in a natural ecosystem, a pest rarely builds its population to a level causing excessive damage, before natural controls come into operation, a quick change in environmental conditions will have a different impact on prey-host population size (Decker et al. 1985). On the other hand, if the change in environment is gradual, the relative response in all organisms could be gradual – so that no change would be noticed unless investigated carefully.

Other factors through which climate change could affect the biomass growth are soil erosion, soil productivity, changes in the cropping system, and moisture regimes. Table 8.2 presents a summary of such impacts.

Table 8.2 Response of crop plants to an increase in CO_2 concentration above current ambient level

Process	Knowns	Uncertainties	Unknowns
Leaf photosynthetic rate	Increase in all plants on first exposure. C_3 responds more than C_4. Response follows law of diminishing returns with little further response after 1,000 (ppm)		Magnitude of response for some crop species
Inhibition of photosynthesis by source and sink imbalance	Response occurs in many species	Response is somewhat correlated with starch accumulation	Mechanism underlying response. Magnitudes of response under various conditions
Leaf transpiration rate	Decrease in all plants. C_4 plants response more than C_3	The reduction in transpiration rate reduces latent heat loss and increase temperature	Stomatal mechanism giving response. Magnitude of response under various environmental conditions
Leaf anatomical and biochemical adaptation	Leaf area, weight per unit area, thickness, and number of mesophyll cell layers increase in many species	RuBP carboxylase activity is reduced in some species. For practical purposes, counteracting responses cancel and adapted, leaves have photosynthetic rates similar to unadapted. Some studies show higher carbon exchange rate for leaves adapted to high CO_2	Mechanism underlying responses
Canopy leaf area	Usually increase		Magnitude of interaction between CO_2 and other environmental and crop variables
Canopy photosynthetic rate	Similar response as leaf photosynthetic rate	Responses can be predicted from a knowledge of how leaf photosynthetic characteristics vary with previous light exposure	Magnitude of response for some crop species
Yield	Very variable depending on harvested parts and on other environmental conditions. Increase 32% on average between 330 and 650 ppm for plant in favorable conditions		Many of the mechanisms that determine yield

(continued)

Table 8.2 (continued)

Process	Knowns	Uncertainties	Unknowns
CO_2 × light interaction	Photosynthetic rate and absolute dry weight gain respond more to high CO_2 in high light in most species, but there is still a small response in low light in C_3 species		Magnitude of response some crop species
CO_2 concentration × temperature interaction	Extremely high or low temperature limits source or sink strength and limit response to CO_2	High CO_2 allows okra to survive low temperature that would otherwise kill it and allows grape to tolerate high temperatures	Magnitude of response of most crop species
CO_2 concentration × soil water availability	Transpiration rate is reduced in high CO_2 (after canopy closure), and plants need less water	Plants maintain high turgor pressure. Relative yield response to CO_2 is greater at low soil water	Many of the mechanisms involved. Magnitude of response
CO_2 concentration × mineral nutrient	Relative response of C_3 species to CO_2 is generally, but not always, less in nitrogen deficiency. In legumes, nitrogen fixation keeps pace with C fixation	Potassium shortage inhibits any response to CO_2. Nitrogen shortage inhibits response to C_4 species to CO_2	Magnitude of response for most crop species
CO_2 concentration × salinity		Relative response to CO_2 is higher in saline condition	Magnitude of response for most crop species
CO_2 concentration × gaseous pollutants	In C_3 species, high CO_2 reduces the effect of pollutant because stomatal conductance is reduced	In C_4 species, SO_2 damage is worse in high CO_2	Mechanism in C_4 species response. Magnitude of response for most crop species and pollutants
Canopy transpiration rate	Decreases after canopy closes. Response before canopy closure depends on counteracting effects of decreased stomatal conductance and increased canopy leaf area		Whether response can be predicted from knowledge of how leaf characteristic vary with depth in canopy. Magnitude of response under various conditions

Carbon partitioning among organs	Proportion of carbon going to roots and stems is increased in many but not all species		Mechanism of response. Magnitude of response under various conditions
Stem growth	Stem height increases in many species	Volume and/or density of wood increases in woody species	Magnitude of response in species where stem is harvested
Root growth		Number of active root axes, and hence, total root length increases	Magnitude of response and whether it affects amount of soil water available to plant
Branching, flowering, and fruiting	Limitation and retention of these organs are increased in many species	Development of organs is generally slightly earlier	Mechanism underlying response. Magnitude of response under various conditions
Canopy water-use efficiency	Increase in C_3 and C_4 plants. Increase in photosynthesis of yield contributes more than reduction in transpiration		Magnitude of response under various conditions
Fruit and seed	Increase number and/or size of fruits and seeds	No significant change in seed analysis	Magnitude of response under various conditions

Source: Acock and Allen (1985)

Acock and Allen (1985) have also mentioned that generally C_4 plants show little response in photosynthesis to increase CO_2 concentration above 340 ppm but show considerable decrease in stomatal conductance – resulting in an increase in photosynthetic WUE (water-use efficiency) – mainly because of a reduction in transpiration. On the other hand, C_3 plants show considerable increase in photosynthesis with doubling of CO_2 concentration, and less decrease in stomatal conductance, which results in an increase in leaf-level WUE attributable more to the increase in photosynthesis than the reduction in transpiration.

8.8 Environmental Conditions and Growth Partitioning

Scientists are almost sure that there is growth enhancement due to increase in CO_2 concentration, but knowing where the growth takes place is important for foresters because they may be interested in the growth of particular organs. There is a definite order of priorities by which plants allocate the carbon, which Acock and Allen (1985) suggested to be in order of:

1. Survival
2. Reproduction
3. Growth of existing organs
4. Increase in the number of organs
5. Storage of access carbon for the future

Where, for some reasons, the higher priorities of carbon allocation cannot be met, the next priority is usually addressed. When all possible uses for carbon are being met, some plants stop fixing carbon. When some material essential to plant growth is in short supply, the organ nearest the source of that material will likely have proportionately more carbon partitioned to them. According to this principle, if a plant grows in a limiting condition of carbon supply, it will preferentially partition carbon to the leaves, which will have the effect of increasing the carbon fixation rate. If the same plant is grown in higher concentration of CO_2, the supply of carbon will be increased, and proportionately, more of the carbon will reach the stems and roots. However, for the same concentration of CO_2, the nature of growth may differ from plant to plant. For example, stem growth may occur without height growth, or leaf area may be increased without any increase in specific leaf area or leaf weight. Hence, other environmental factors control, in most cases, what would be the nature of the growth.

We can assume that as the CO_2 concentration is increasing, so does the concentration of associated atmospheric pollutants. Since most gaseous pollutants enter the leaf via the stomata and because high CO_2 concentrations reduce the stomatal conductance, high CO_2 will probably reduce the damage done by the pollutant. Acock and Allen (1985) mentioned that the presence of 0.25 ppm of atmospheric SO_2 (sulfur dioxide) reduces the growth of several C_3 plant at 300 ppm CO_2 concentration, but not at 600 or 1,200 ppm. The reverse was true for C_4

plants, where the presence of SO_2 reduces the growth in high CO_2 concentration but not in 300 ppm CO_2. This difference in growth was identified in the case of leaf area expansion and dry weight gain. But it is not clear what the relation between photosynthetic pathway (C_3, C_4) and CO_2 concentration is.

Higher CO_2 concentrations may help plants to grow in adverse conditions. For example, in saline areas where stress from moderate salinity maintains higher respiration rates, plants probably have to use the additional energy to keep the salt away from the tissue. Growing plants in higher CO_2 concentration may meet this demand. So, where there could be salinity intrusion, the higher CO_2 environment may make the plants relatively more salt tolerant (Table 8.2).

8.9   Derived Effects of CO_2

The total integrated effects of elevated CO_2 on ecosystems could be extremely complex and hence are very difficult to study. The community and ecosystem factors of prime importance are primary production, carbon accumulation, plant species composition, water use, animal abundance, and animal diversity. Important effects of elevated CO_2 on nutrient cycling, herbivory, decomposition, and mineralization are also expected.

8.9.1   Nutrient Availability

Elevated CO_2 and concomitant increase in growth potential would be expected to deplete and sequester the limited nutrient reserves of many ecosystems. The limits to nutrient availability may lead to an increase in the C/N ratio of litter leading to lower rates of decomposition, nutrient turnover, and mineralization. On the other hand, some responses to elevated CO_2 may act to increase nutrient availability. The circumstances of increased carbohydrate availability to decomposer and symbionts may be such that symbiotic nitrogen-fixing organisms and mycorrhizal associates will act to counterbalance the decrease in nutrient availability (Lamborg et al. 1983). Increased transport of carbohydrate to the roots may increase nitrogen fixation through symbiotic nitrogen-fixing organisms.

8.9.2   Carbon Sequestering

With an elevation in atmospheric CO_2, there may be an increase in the productivity of ecosystem, and hence, carbon sequestering may be affected in several ways. Net primary productivity (NPP) may be increased. An initial increase in NPP should increase litter fall more than decomposition, and this process may continue until equilibrium is reached. But where atmospheric CO$_2$ rises at a gradual level of 1–1.5% per year, rates of increase of NPP and litter fall may be more modest. In this case, litter fall will increase at a rate less than 1.5% annually. But plants in mature ecosystems have a near energy balance of gross primary production (GPP) and respiration. In this situation, plant growth has to be slowed because the support of a large respiring biomass utilizes nearly all available photosynthates. If the increase in CO$_2$ is a rapid step increase, then plants could grow quickly until maintained respiration is again equaled by photosynthesis. Plant growth and certain plant parameters may respond to the elevated CO$_2$ in a manner similar to the ecosystems. LaMarche et al. (1984) reported a 100% increase in tree ring width with a 25% increase in CO$_2$ concentration. If the carbon available for wood production could be drawn from the residual between the net photosynthesis and dark plant respiration, then a doubling of atmospheric CO$_2$ might increase net photosynthesis only 10–30% while increasing the carbon available for ring production by 100% or more. But higher growth rates may not result in a greater standing crop if harvested early.

8.9.3 Plant Animal Interaction

Plant animal interaction is important in the development and distribution of species in the ecosystem. The relationship may be changed due to changes in the growth pattern and the biochemical contents of the components. In the case of plant community, the derived changes in the plant animal interaction can be described as follows:

8.9.3.1 Pollination and Seed Dispersal

CO$_2$-induced changes in the geographic distribution of plant species have been postulated by Oechel and Strain (1985). These postulations are based on changes in environmental temperature, precipitations or water availability, and water use or in the tolerance of plants to the environmental stress at elevated CO$_2$. As the range of plants expands into the new areas, the possibility exists that the appropriate pollinators may not or cannot move with the plants. As a result, pollinators may not exist in the new habitat. In addition, the activities of plants in the new environment may be out of phase with the animals required for pollination or seed dispersal.

8.9.3.2 Herbivory

The inorganic and organic quality of plant tissue is likely to change undergrowth at elevated CO$_2$. However, the exact nature of the change will depend on both the environment and the species involved. Likely outcomes include plant materials with

Table 8.3 Effects of competition and CO_2 concentrations on total dry weight, leaf area, and height of soybean and Johnsongrass after 24 days

Species	Competitor	Dry weight (g) CO_2 conc. (ppm)		Leaf area (dm²) CO_2 conc. (ppm)		Height (cm) CO_2 conc. (ppm)	
		350	675	350	675	350	675
Soybean (C_3)	None	6.92	10.43	14.68	16.20	53.6	53.9
	Soybean	5.00	6.80	10.23	10.50	51.5	53.4
	Johnsongrass	5.40	8.15	11.69	12.74	51.2	52.8
Johnsongrass (C_4)	None	5.62	6.19	10.39	10.68	90.6	106.4
	Soybean	3.72	3.70	7.43	5.75	88.3	98.6
	Johnsongrass	4.23	5.20	8.00	7.82	94.2	102.5

Source: Patterson et al. (1984)

lower inorganic nutrient content. Certain herbivores will adjust to this lower nutrient content by increasing the consumption. This suggests that greater productivity of plants at elevated CO_2 may be offset in plant by increased herbivory.

8.9.4 Plant-Plant Competition

In the field condition, there may be a number of species growing together. Their strategies may change under the condition of elevated CO_2. Bazzaz et al. (1985) suggested that an elevated CO_2 level will affect the competitive interaction of C_3 and C_4 species (Table 4.2). Carter and Peterson (1983) grew *Festuca elatior* (C_3) and *Sorghum halepense* (C_4) in mixed and unmixed culture under 350 and 600 ppm CO_2 for 112 days. These have been used to project that global CO_2 enrichment might alter the competitive balance between C_3 and C_4 plants and subsequently affect seasonal niche separation, species distribution patterns, and net primary production within mixed communities. Patterson et al. (1984) studied the growth of a weed Johnsongrass (*Sorghum halepense*) (C_4) and a soybean crop (*Glycine max*) (C_3) and their interaction under 350 and 675 ppm CO_2 concentration. The results are given in Table 8.3.

In the absence of competition, the CO_2 concentration stimulated higher dry weight accumulation, leaf area expansion, NAR (net assimilation rate = dry matter production per unit leaf area) and LAD (leaf area duration = photosynthetic rate × leaf area × time) of soybean more than that of Johnsongrass. The plant relative yield (yield in mixture/yield in pure culture) of Soybean in competition with Johnsongrass increased and that of Johnsongrass in competition with soybean decreased, as the CO_2 concentration was increased from 350 to 675 ppm. Therefore, the competitiveness of the C_3 crop with C_4 weed increased with an increase in CO_2 concentration. Bazzaz and Carlson (1984) studied the competition among four species, three C_3 and one C_4, at two levels of moisture and reported that the effects of moisture and CO_2 concentration amplified each other. Total biomass of the community increased, but the competition was limited between C_4 plants and only one of the C_4 plants. However, both investigations can be considered as preliminary due to limited replication and the relatively short duration of the experiment.

Zangrel and Bazzaz (1984) investigated the growth of six members of an annual community to combinations of CO_2 and nutrient concentration and CO_2 concentrations and light intensities until reproductive maturity. He found that under low light levels, total community production was below the maximium at the highest CO_2 concentration. This suggests that some environmental conditions may cause the production of some members of the community to decrease with elevated CO_2 concentrations. Once again, acceptance of the result is affected by limited replication in the experimental design.

References

Acock B, Allen LH (1985) Crop responses to elevated carbon dioxide concentrations. In: Strain BR, Cure JD (eds) Direct effects of increasing carbon dioxide on vegetation. US Department of Energy, Washington, DC, pp 53–97

Agrois GN (1978) Plant pathology, 3rd edn. Academic, New York

Aston AR (1984) The effect of doubling atmospheric CO_2 on stream flow: a simulation. J Hydrol 67:273–280

Bazzaz FA, Carlson RW (1984) The response of plants to elevated CO_2 competition among an assemblage of annuals at different levels of soil moisture. Oecologia 62:196–198

Bazzaz FA, Garbutt K, Williams WE (1985) Effect of increases of atmospheric carbon dioxide concentration on plant communities. In: Strain BR, Cure JD (eds) Direct effects of increasing carbon dioxide on vegetation. US Department of Energy, Washington, DC, pp 155–170

Callaway JM, Currie JW (1985) Water resources systems and changes in climate and vegetation. In: White MR (ed) Characterization of information requirements for studies of CO_2 effects: water resources, agriculture, fisheries, forests and human health. DOE/ER-0236. United States Department of Energy, Washington, DC, pp 24–67

Carter DR, Peterson KM (1983) Effects of CO_2 enriched atmosphere on the growth and competitive interaction of a C_3 and C_4 grass. Oecologia 58:188–193

Cremer KW (1976) Daily patterns of shoot elongation in *P. radiate* and *E. regnans*. New Phytol 76:459–468

Decker WL, Jones V, Achutuni R (1985) The impact of CO_2-induced climate change on U.S. agriculture. In: White MR (ed) Characterization of information requirements for studies of CO_2 effects: water resources, agricultures, fisheries, forests and human health. DOE/ER-0236. United States Department of Energy, Washington, DC, pp 71–91

Gholz HL, Ewel KC, Teseky RO (1990) Water and forest productivity. For Ecol Manage 30:1–18

Iso SB, Brazel AJ (1984) Rising atmospheric carbon dioxide concentration may increase carbon dioxide on vegetation. US Department of Energy, Washington, DC, pp 185–204

Klepper B (1968) Diurnal pattern of water potential in woody plants. Plant Physiol 43:1931–1934

Kramer PJ, Kozlowski TT (1979) Physiology of trees. McGraw-Hill, New York

LaMarche VC Jr, Graybill DA, Fritts HC, Rose MR (1984) Increasing atmospheric carbon dioxide: tree ring evidence for growth enhancement in natural vegetation. Science 225:1019–1021

Lamborg MR, Hardy RWF, Paul EA (1983) Microbial effects. In: Lemon ER (ed) CO_2 and plants: the response of plants to rising levels of atmospheric CO_2. Westview Press, Boulder, pp 131–176

Luxmoore RJ (1991) A source sink framework for coupling water, carbon, and nutrient dynamics of vegetation. In: Kaufmann MR, Landsberg JJ (eds) Advancing towards closed models of forest ecosystems. Heron Publishing, Victoria. A special volume of Tree Physiol 9:267–280

Oechel WC, Strain BR (1985) Native species responses to increased atmospheric carbon dioxide concentration. In: Strain BR, Cure JD (eds) Direct effects of increasing carbon dioxide on vegetation. US Department of Energy, Washington, DC, pp 117–154

Patterson DT, Flint EP, Beyers JL (1984) Effects of CO_2 enrichment on inter-specific and intra-specific competition between a C_3 crop and a C_4 weed. Weed Sci 32:101–105

Solomon AM, West DC (1985) Potential responses of forests to CO_2-induced climate change. In: White MR (ed) Characterization of informal requirements for studies of CO_2 effects: water resources, agricultures, fisheries, forests and human health. DOE/ER-0236. United States Department of Energy, Washington, DC, pp 146–169

Walker BH, Young MD, Cocks MCR, Landsberg JJ, Fleming PM, Parslow TS (1990) Global climate change and Australia – effects on renewable natural resources. In: Swine DJ (ed) Greenhouse and energy. CSIRO, Melbourne, pp 24–27

Zangrel AR, Bazzaz FA (1984) The response of plants to elevated CO_2 – competitive interactions between annual plants under varying light and nutrients. Oecolgia 62:412–417

Chapter 9
Perspective of Climate Change in Bangladesh

There are arguments both for and against the effects of change in greenhouse gases on climate. If the assumption of a climatic effect is valid, the change in the regional climate may have some impact on the socioeconomic condition, and this may be particularly acute in some developing countries such as Bangladesh. The socioeconomic impact may also cover social health of the people. As there is some likelihood of a significant impact, strategic measures to prevent or control the greenhouse effect are becoming more important day by day. The importance of these strategic measures varies from society to society depending on their socioeconomic conditions. For example, given the dense population and poor social conditions in Bangladesh, there will be less awareness of the greenhouse effect than in Australia. The study of socioeconomic development is an important factor in strategic planning to combat the climatic impact of a single nation (several global organizations are already working on this issue). In order to improve the socioeconomic conditions of poor countries, a unified global action is needed so that the measures to combat the climate change effect do not appear as a hindrance to the development of such nations.

9.1 Emission Problems Under the Socioeconomic Situation of Bangladesh

Before explaining the strategic measures for controlling climate change situations in a particular society, we have to know the major contributing gases produced under the socioeconomic conditions of that country. Gases other than CO_2 are thought to be responsible for about one-third of present global warming, and it is estimated that these will cause about half the problems associated with the greenhouse effect by around 2030 (Anon 1987). In Sect. 3.2.3, it was mentioned that denitrification is most often observed in environments isolated from atmospheric oxygen and supplied with adequate oxidizable detrital materials, organic rich sediments, flooded soils, and closed oceans basins (McElroy and Wofsy 1988).

M. Ali, *Climate Change Impacts on Plant Biomass Growth*,
DOI 10.1007/978-94-007-5370-9_9,
© Springer Science+Business Media Dordrecht 2013

Table 9.3 Global primary energy consumption 1984

World Bank GNP economy category	GNP per capita (1984 dollars)	Energy use (kW/capita per year)	Population in 1984 (million)	Total consumption TW[a]
Low income	260	0.14	2,390	0.99
Sub Sahara	210	0.08	258	0.02
Middle income	1,250	1.07	1,188	1.27
Lower middle	740	0.57	691	0.39
Upper middle	1,950	1.76	497	0.87
Sub Sahara	680	0.25	148	0.04
High income oil exports	11,250	5.17	19	0.10
Industrial market econ	11,430	7.01	733	5.14
E. Europe nonmarket econ	–	6.37	389	2.44
World	–	2.11	4,718	9.94

Source: Anon (1987)[a]TW is terawatt = one billion kW

one vehicle to each person, contributing CH_4 and other components of greenhouse gases many times greater than the poor people in developing countries. Energy use standards reveal how devastating the impact of people's habits can be on their environment and the extent to which such devastation is dependent on the state economy.

Table 9.3 shows the relation of energy use to the corresponding GNP. This table gives an idea of the range of future energy needs from which quantity of future greenhouse gas emission could be predicted. If the energy consumption per head were to become uniform worldwide at current industrial country levels by 2025, some global population would require about 35 terawatts (TW). In more conservative estimates, if the low and middle economies need 10.5 TW, the total global energy needs in the future would be 20 TW (Anon 1987). More energy use means more greenhouse gas emission. If such an increase in energy use were to happen in Bangladesh, 158 million people would be responsible for about 1.3 TW of greenhouse gases from the energy-use sector. At present, Bangladesh consumes less than 0.04 TW (calculated from Table 9.2) of energy per year, and the major source of that energy is biomass.

Biomass has been used from historical times as a source of energy. But, it seems that the present increase in greenhouse gas emission has resulted from an increase in the use of fossil fuels rather than biomass fuel as an energy source. Because emission from biomass fuel is again absorbed by the new growing biomass, it therefore maintains a balance. However, emissions coming from industrial fuel are emissions we are making from burning oils on a single day that has been stored as fossil fuel below the ground for millions of years.

Over the past century, the use of fuel has grown nearly 30-fold, and industrial production has increased 50-fold. About 75% of the present day increase in greenhouse gas concentration in the atmosphere results from the increase in the use of fossil fuel, and over 80% in the increase in industrial production is included. The annual increase in global industrial production today is perhaps as large as the total production of Europe around the end of the 1930s (Anon 1987). The consequences of such an increase in emission will be serious for countries like Bangladesh despite their undeveloped status.

Although no major study has been done so far, there are many reports from Bangladesh about the pollution of river water, especially in the industrial towns due to dumping of noxious wastes and effluents from the banks of the rivers of Bangladesh (Choudhury 1985). It seems that when countries like Bangladesh attempt to adapt to accelerate industrial development, they will lose control not only of greenhouse emissions but also the spreading of other pollutants. If we take account of what is happening in parts of Europe and North America due to industrial pollution, the future consequences for Bangladesh are partially clear. Due to the industrial revolution, a variety of air pollutants are killing trees, destroying lakes, and damaging buildings and cultural treasures close to, and sometimes thousands and miles from, the point of emission. Acidification of the environment threatens large areas of Europe and North America. Central Europe is currently receiving more than 1 g of sulfur on every square meter of ground each year (Anon 1987). If this also happens in Bangladesh, the loss of forests might not only cause more severe greenhouse effects more quickly but could also bring in its wake disastrous erosion, siltation, floods, and local climate change.

9.2 Outline of Climate Change-Related Problems of Bangladesh

In addition to the local climate change, there could be a global climate change due to the greenhouse effect. A prediction of global climate change is that polar temperatures will probably increase at a rate two or three times greater than at the equilateral zone (Anon 1987). The change in the temperature regime will, in turn, change the dynamics and water-holding capacity of the atmosphere resulting in a change in the rainfall regime. General agreement is lacking on how increased CO_2 levels will influence the amount of precipitation.

Decker et al. (1985) predicted the following:

1. The increase in precipitation will be negligible in the lower latitudes.
2. The precipitation will show a slight increase between the latitudes 15° and 35 °N.
3. The precipitation will actually decrease in the latitudes between 37° and 47 °N.
4. There will be a substantial increase of precipitation in the higher latitudes because of greater poleward transfer of water vapor. The evaporation rate, however, he suggested, would increase uniformly over all latitudes.

Bangladesh is situated in the tropical zone with latitude boundaries at 20.34 °N and 26.38 °N and longitudinal boundaries at 88.01 °E and 92.41 °E (Fig. 9.1). The country is characterized by a high-temperature humid monsoon climate. Bangladesh

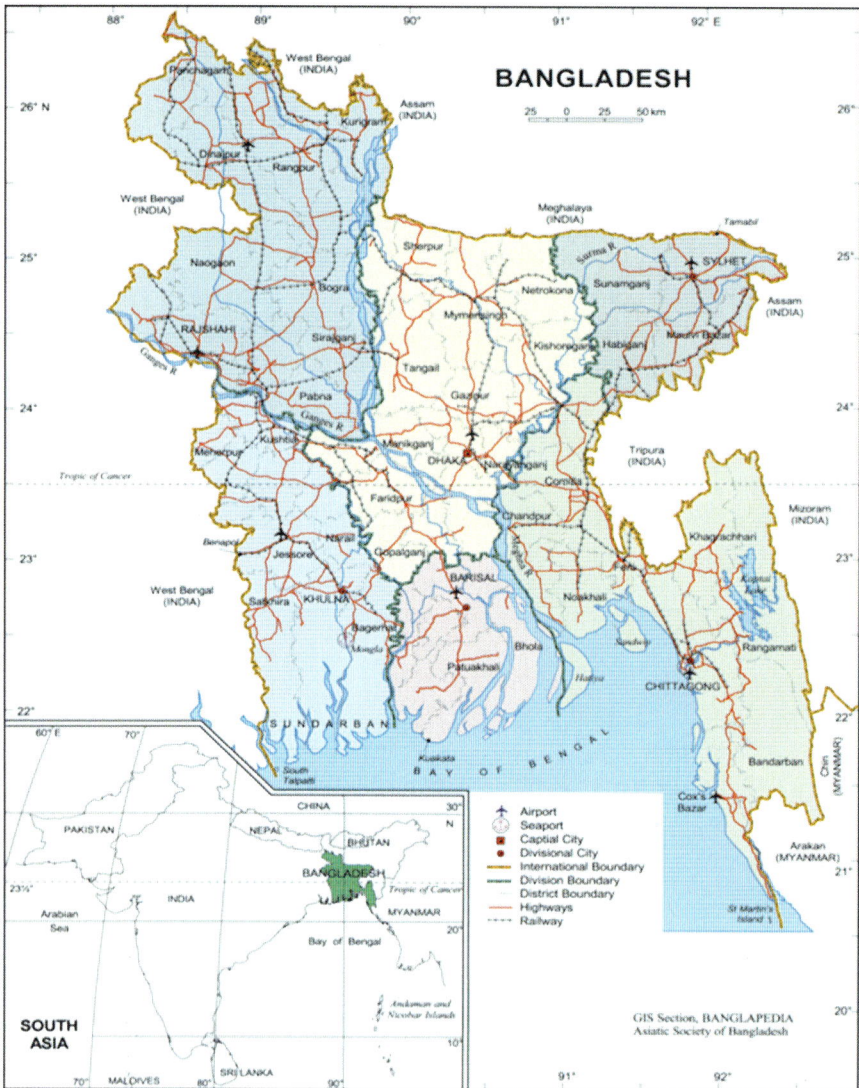

Fig. 9.1 Map of Bangladesh showing the geographical position and location within the subcontinent

is a low-lying riverine country of about 144,000 km^2 (Choudhury 1985) formed by the rivers Ganges, Brahmaputra, Meghna, Karnaphuli, and Sangu (Fig. 9.2). More than 85% of the country is flat alluvial plain crisscrossed by the rivers and their innumerable tributaries and distributaries. These rivers are an important source of irrigation and abundant supply of fish. The network of rivers and their tributaries, numbering about 230 with a total length of about 15 million miles covering the

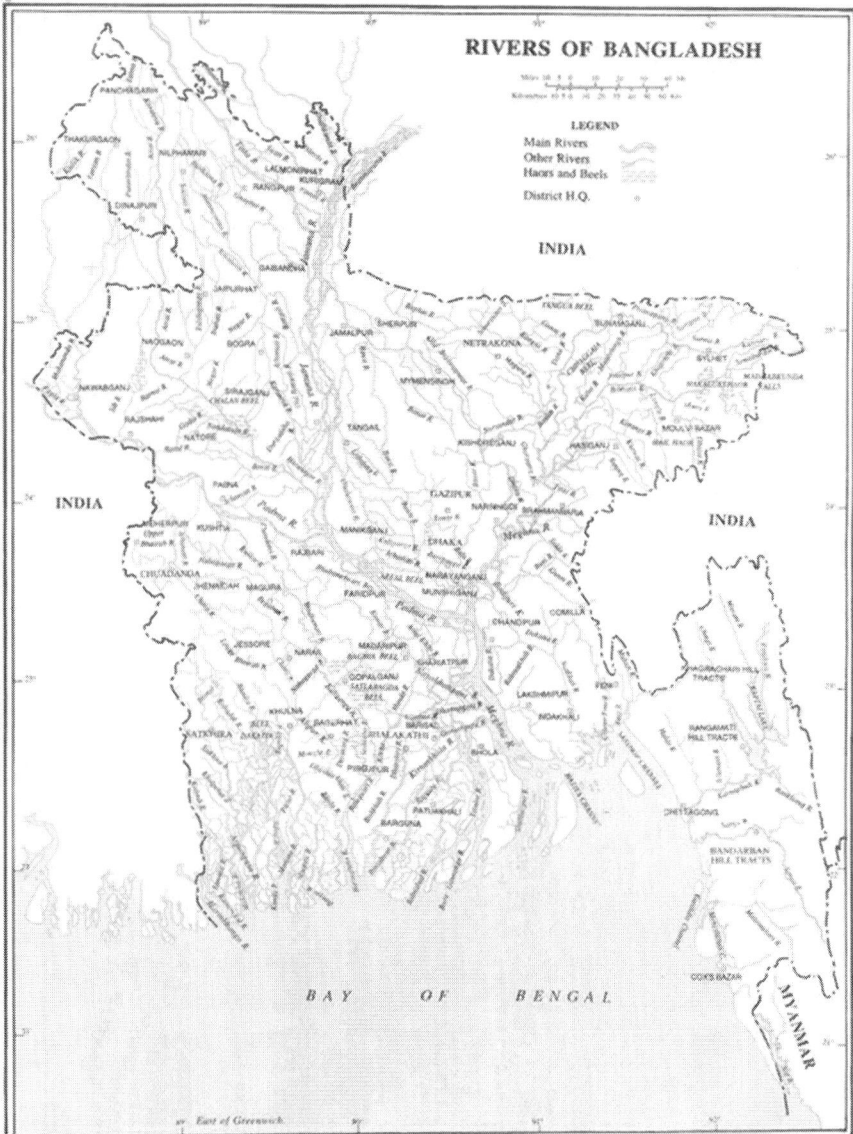

Fig. 9.2 Map of Bangladesh showing the river network, relief and anticipated vulnerability to climate change impact

entire country, flow down to the Bay of Bengal (Fig. 9.2). In 1970 (pre-Farakkha Barrage period), the amount of surface water that normally flowed through was estimated at 718 million acre feet (maf) per year. Out of these, 708 maf passed through the Ganges-Brahmaputra Delta comprising an area of about 2,500 sq miles.

Unfortunately, because of geographical location surrounded by India (Fig. 9.1), Bangladesh is also affected by what is happening in India. India is now the world's 4th largest polluter, achieved through her rapid economic growth. Das (2009) reported that Orissa state's power stations in India alone emitted 164 million tons of carbon dioxide in 2005, equal to the total emitted in the whole of India in 1996, and it will emit 3% of global emissions after new planned plants are opened. All these shoots are depositing at the Himalaya and causing early melting of ice cap. The additional melted water is draining through Bangladesh, the down floodplain, and is causing problems now and will cause future problems through changing water availability.

Within Bangladesh, as in the region (2) classified above, a precipitation increase along with an expected sea level rise due to the greenhouse effect could create several economic and environmental problems. Agricultural activities, the mainstay of the Bangladesh economy, would receive a severe setback along with forest and general ecological destruction. Perhaps there will be no vegetation left in Bangladesh to receive the benefit of enhanced growth. The severity and frequency of floods and storms could increase and destroy things. Therefore, the enhanced growth, what we are talking about, will depend on the area. One feature of the scenarios formulated is that any CO_2-induced climate change would be gradual and the CO_2 level could be stabilized, and new climate equilibrium would emerge – perhaps without countries like Maldives and Bangladesh.

It is often said that agricultural production would be more affected by climate variability than by climate change. In tropical regions like Bangladesh, even small amounts of warming will lead to declines in the amount of crops harvested. Agricultural output in developing countries is expected to decline by 10–20% by 2080, and temperature increases of more than 3°C may cause food prices to increase by up to 40% (Cohen et al. 2008). The agricultural system can adapt if the climate variability is low and gradual, and, hence, the climate associated with a new equilibrium may have its own crop pattern. But high regional climate variability may occur in the long interim period of climate change well before an equilibrium is reached, which might not only destroy the existing crop pattern but could also delay the achievement of an equilibrium pattern for decades or even centuries. Bangladesh could lose its agriculture well before an equilibrium climate is reached, mainly due to loss of land under water and salinity, resulting from the expected consequences of greenhouse effects. Allison et al. (2009) predicted from his study on fisheries resource impact on 132 countries that Bangladesh will be one of the worst affected country.

Presently there are climate variabilities, and frequent extreme climates are already appearing to be destructive to agricultural production of Bangladesh. Interannual variabilities are of much greater magnitude than either the natural or CO_2-induced climate change expected over an extended number of years. Tropical cyclones like SIDR and AYELA can flatten the whole country and kill thousands. Probably the severity and frequency of year to year climate variability in Bangladesh forecasts that an induced change is imminent.

Some other environmental factors affecting agriculture are weather events of short durations such as hailstorms, droughts, and periods with above-normal rainfall persisting for days or weeks. Still others may be associated with long-term climate changes, including changes in the frequency of climate events which are important to agriculture, for example, changes in the frequency of major droughts, flooding, temperature extremes, or climate-induced diseases in agriculture production. If higher concentration of CO_2 in the atmosphere causes significant changes to any of these factors, agriculture will certainly be affected, and these effects in case of Bangladesh would always be severe.

The variability of weather, both in terms of temperature and precipitation, in a given area or regions is influenced primarily by the location of the ridges and the troughs in the tropospheric circumpolar circulation (jet stream) (Decker et al. 1985). Changes in the regional climates due to changes in the location frequency of the circulation pattern will, in the short term, overwhelm any gradual drift in the mean annual and seasonal temperatures. Thus, variability as a climatic risk is much greater for agriculture than a gradual warming of, say, 1–3°C. So, high regional climate variability due to an increase in atmospheric CO_2 concentration would be more critical for agricultural production than slow global climate change.

For any given agricultural region, the new climate may induce greater or smaller climate risk depending on the relief features of that region. The relief features of Bangladesh are very susceptible to a sea level rise due to greenhouse effects. Most of the country is a plain giving the impression of absolute flatness except for a few abrupt areas of old alluvium elevated to about 30 m. An increase in 3 m in sea level would engulf the whole plain land, 80% of the total area of the country (Ahmad 1976). If the rise in the sea level is 1 m, it will engulf one-third of the country. Because most of the agricultural land of Bangladesh is climatically marginal and agriculturally unsustainable, therefore, agriculture over a vast area will not be possible due to loss of land under water and intrusion of salinity. In this one-third inundated flat land, the Sundarbans, a mangrove forest area of 2,350 square miles, would also be included. The displaced could move to the 20% hilly lands, and they would destroy the remnants of vegetation. Hence, Bangladesh would lose its whole vegetation and the territorial characteristics needed to form a state. At present, agriculture in Bangladesh is so dependent on climate that the predicted change in climate would affect both the magnitude and efficiency of the production system.

The economy of Bangladesh mainly depends on agriculture (about 65% according to Decker et al. 1985). Although the manufacturing sector (garments) and foreign remittance have overtaken, still 30% of Bangladesh GDP and 60% employment are coming from agriculture. A change in agriculture production due to greenhouse effects will affect directly 75% of the population (about 114 million) who depend on the agriculture and others indirectly through reduction of nation's gross domestic products (GDP) from the agricultural sector. About 56% of the GDP of Bangladesh is from agriculture, and 60% of the industrial output depends on agricultural raw materials. The production of these industries will be affected, which means that another 10% of GDP will be affected indirectly (Ahmad 1976).

Hence, a reduction in the biomass production will affect the economy of Bangladesh drastically.

If the worst case of the climate change is considered, the consequence for Bangladesh could be fatal – a country which will probably be totally wiped from the map of the world. But as the process will be very gradual, a definite forecast is yet impossible. Continued growth in the population of Bangladesh suggests that, despite the best efforts, hunger and malnutrition will be prevalent in the twenty-first century with or without changes in the climate. Whatever the effects of greenhouse gases elsewhere, there is only scarce green area (forest) left, only 6% of the total country; therefore, positive effects, if they occur, will not be visible.

9.3 What Has Been Done

An important step toward environmental protection was taken in 1977 with the promulgation of the Environment Pollution Control Ordinance (Ordinance No XIII of 1977), which provided for the control, prevention, and abatement of pollution of the environment. This ordinance deals with regulations of industrial discharge but not something like CO_2 or CFC emissions in the atmosphere. In fact, it would be economically very hard for Bangladesh to cut off CO_2 or CFC emissions in the atmosphere.

Under this ordinance, an appointed director will ensure the implementation of this ordinance. The Pourashava authorities (towns and cities), Zilla Parishad, Upazila Parishad, and Union Parishad have been given responsibility for management of the environment within their respective jurisdictions (Choudhury 1985). In practice, the authorities of these organizations have not yet developed programs to protect and preserve a healthy and natural environment in the towns or countryside. Gain (2002) has reported that Bangladesh has about 30,000 industrial units: 24,000 small- and cottage-sized and 6,000 large- and medium-sized factories. These industries have a very poor environmental record. As Bangladesh has experienced one of the fastest industrial growth rates in South Asia and their numbers are growing, there must be pressure from civil society directly and through government to improve their energy efficiency, dispose of waste properly, and to reduce emissions. The funding probably was inadequate. The planning of laws and regulations should provide ample scope for laying down standards intended to provide environmental quality. However, one positive initiative taken by the Bangladesh government is the availability of loans from the Bangladesh Bank to help brick kilns reduce their emissions by using *Hybrid Hoffman Kilns* or similar technology that can cut carbon dioxide emissions by 5,000 tons per year per unit. This is important as brickfields in Bangladesh emit around 87,500,000 tons of carbon dioxide per year (Star Business Report 2010).

The Bangladesh government has already committed itself to achieving carbon neutrality (Rahman 2009). Bangladesh is convinced that reducing dependence on carbon in economic growth and promoting climate resilience will promote both development and poverty eradication in a sustainable manner (SAARC 2010) and has promised to prevent methane (a powerful greenhouse gas) being released during coal mining and to using clean coal technology in power stations (MoEF 2008). So there are plenty of commitments that the government can be encouraged to honor and implement through advocacy action.

Energy use is another way greenhouse gas emissions can be reduced in Bangladesh. Natural gas produces 24% of the country's fuel need and imported coal and mineral oil 19% (Moral 2002). The latter produces large quantities of greenhouse gases to produce electricity, while although very much less, gas still produces some carbon dioxide when burnt. Due to power shortages in Bangladesh and regular "load shedding," there is a lot of pressure to follow India and China and to expand coal mining and build more coal-fired power stations. Bangladesh must not follow the same damaging route, although the proposal makes short-term economic sense. A matter for hope is that Bangladesh is thinking of a nuclear power station for supplying her energy needs.

References

Ahmad N (1976) A new economic geography of Bangladesh. Vikas Publishing House Pvt. Ltd., New Delhi

Allison EH, Perry AL, Badjeck MC, Adger WN, Brown K, Conway D, Halls AS, Pilling GM, Reynolds JD, Neil L, Andrew NL, Dulvy NK (2009) Vulnerability of national economies to the impacts of climate change on fisheries. Fish Fisheries 1–4

Anon (1987) Our common future. World Commission on Environment and Development (Oxford University Press)

Choudhury AKMK (1985) Land use planning in Bangladesh. National Institute of Local Government, Dhaka

Cofer WR, Levine JS, Sebacher DI, Winstead EL, Riggan PJ, Stocks BJ, Brass JA, Ambrosia VG, Boston PJ (1989) Trace gas emissions from chaparral and boreal forest fires. Great Lake Forestry Centre

Cohen MJ, Tirado C, Aberman N, Thompson B (2008) Impact of climate change and bioenergy on nutrition. Food & Agriculture Organisation of the United Nations (FAO) and International Food Policy Research Institute (IFPRI), Rome, 86 pp

Crutzen PJ, Heidt LE, Krasnec JP, Pollock WH, Seiler W (1979) Biomass burning as a source of atmospheric gases CO, H_2, N_2O, NO, CH_3Cl and COS. Nature 282:253–256

Das A (2009) Integrated farming: a climate change adaptation strategy for small and marginal farmers in low lands of sundarbans. 3rd international conference on community based adaptation to climate change. Organised by Bangladesh Centre for Advanced Studies, IIED & The Ring. Radision Water Garden Hotel, Dhaka, 18th – 24th Feb 2009

Decker WL, Jones V, Achutuni R (1985) The impact of CO_2-induced climate change on U.S. agriculture. In: White MR (ed) Characterization of information requirements for studies of CO_2 effects: water resources, agricultures, fisheries, forests and human health. DOE/ER-0236. United States Department of Energy, Washington, DC, pp 71–91

Gain P (2002) Bangladesh environment: facing the 21st century. Society for Environment and Human Development (SEHD), Dhaka, pp 181–182

McElroy MB, Wofsy SC (1988) Tropical forests: interaction with the atmosphere. In: Prance GT (ed) Tropical rainforest and the world atmosphere. International Book Distributors, Dehradun, pp 33–60

MoEF (2008) Bangladesh climate change strategy and action plan 2008. Ministry of Environment and Forests, Government of the People's Republic of Bangladesh

Moral S (2002) Energy resources. In: Gain P (ed) Bangladesh environment: facing the 21st century. Society for Environment and Human Development (SEHD), Dhaka, pp 162–180

Rahman AA (2009) Seal the deal in Copenhagen: the most vulnerable communities demand climate justice. Clime Asia. COP 15/MOP5 Special Issue, 1–8

SAARC (2010) Thimphu statement on climate change. Sixteenth SAARC Summit. Thimphu, Bhutan, 28–29–April 2010

Star Business Report (2010) Brick kilns to get bank loan to go green. The Daily Star. Dhaka, April 9, 2010

van den Broeck B, Sinanian A (1990) Landfill gas management. In: Swine DJ (ed) Greenhouse and energy. CSIRO, Melbourne, pp 157–164

Williams DJ (1990) Australian methane fluxes. In: Swaine DJ (ed) Greenhouses and energy. CSIRO, Melbourne, pp 165–176

Chapter 10
Conclusions

Climate change effects are complex phenomena involving water, soil, atmospheric gases, sunlight, and heat. An increase in the concentration of certain gases in the atmosphere causes the greenhouse effect. However, it has been established that there are sinks as well as sources for those gases in the atmosphere. The greenhouse gases may react with each other and with other gases in the atmosphere and partially nullify themselves. Scientists were divided on whether the greenhouse effect will actually happen or not, though on a short-term basis, it is certain that the concentration of different greenhouse gases in the atmosphere is increasing. This increase in greenhouse gas concentration could bring some changes in the growth of vegetation, and if the greenhouse effect develops and strengthens, climate change could bring about a drastic change in the growth and distribution of vegetation.

Among the greenhouse gases, CO_2 is an essential compound for life on this planet. Plants absorb CO_2 from the atmosphere and produce carbohydrates through photosynthesis: these carbohydrates become the food supply for other organisms in the ecosystem. Hence, the CO_2 concentration in the atmosphere influences plant growth as well as other forms of biomass in the system. It is critical to understand how the world's vegetation will respond as the CO_2 concentration of the global atmosphere continues to increase. Although there is considerable uncertainty, estimates of growth enhancement range from 0.5 to 2.0% for each 10 ppm increase in atmospheric CO_2 (Anon 1985). Thus, it is possible that some increase in agricultural yield that has occurred in this century is partly due to increased atmospheric CO_2 concentration.

Not all species are likely to respond to CO_2 enhancement in a similar way. Plants with high conductance for the diffusion of CO_2 would have greater growth than plants with lower CO_2 conductance. Because many agricultural weeds have high conductance, they could have a comparatively larger growth response to increased CO_2 than some desirable crop species with a lower conductance. Such differential growth could bring about changes in the competitive relationships of species of the community, as a result of which community structure and abundance could be affected.

M. Ali, *Climate Change Impacts on Plant Biomass Growth*,
DOI 10.1007/978-94-007-5370-9_10,
© Springer Science+Business Media Dordrecht 2013

Though the growth of plants may be enhanced by an increase in atmospheric CO_2, the food quality of some of the tissues of plants could decline due to presence of excess carbon and relatively less nitrogen. So, an insect pest feeding on biochemically changed leaves may either consume more leaves to meet its food requirements or change its food behavior for survival. Hence, though productivity may increase in the future, there could be an increase in insect predation and higher weed growth. This could jeopardize any increase in biomass growth. However, a number of secondary responses may occur in conjunction with an increased rate of CO_2 fixation. These include changing interactions among competing plants, altered populations of animals, and greater incidence of diseases. This may change the net carbon content of the ecosystem. Thus, the hypothetical carbon-sequestering potential of this system may not be realized.

Studies and observations on the influence of CO_2 on plant growth are not yet complete. Most of the studies reviewed in this book were on seedlings within growth chambers, the result of which may not be applied in the field. Much more information is yet to come. From the scattered information and speculation reviewed in this book, the following comments might be made:

- The response of annuals and perennials to increased CO_2 may be different in respect to the manner, magnitude, and rate of the response. But in a balanced ecosystem with animals feeding on plants, pathogens operating to restrict growth, and plants competing for limiting water and nutrients, it is uncertain whether biomass production in the ecosystem will increase as a whole. Though there might be some biomass increase, community composition perhaps might be changed due to the differential responses of different species and the interaction of dominancy. The life cycle of most plant species might be accelerated. Changing rates of flower, fruit, and seed production may affect coadapted animal consumers and pollinators. Some authors have suggested that reproductive potential may be changed with concomitant changes in gene frequency.
- Increased atmospheric CO_2 may induce nutrient impoverishment in the soil of some ecosystems by accumulation of minerals in the larger plant bodies. If tissues of annuals become nutrient poor, decomposition rates may decline, thus increasing the tendency for minerals to remain sequestered organically.
- Changes in vegetation growth as well as climate change will occur throughout the world. Temporarily, it could be presumed that some of the nations, especially temperate countries, would benefit from these changes. However, this may not be the case because all the changes will not occur at the same time within the one sequence. For example, it is supposed that vegetation will develop in growth as well as in diversity in temperate areas, but this may not happen because the soil condition will not be altered within that limited period. Hence, an intermediate situation, where climatic variables favor one kind of vegetation and soil variables favor another, will jeopardize the situation.
- If a carbon dioxide-induced change is a genuine threat to a harmonious future, then a cheap, albeit partial solution could be provided by increased afforestation. At the same time, there would have to be change to the wood utilization sector

so that C is "locked up" for a long time. Even if climate change proves to be unimportant physically or politically, foresters should still address the issues of "aerial fertilization" and wood utilization in their management planning and prescriptions because superior tree families under existing levels of CO_2 will not be necessarily superior under doubled CO_2 levels.

The ultimate consequences of the greenhouse effect remain unknown. Though some nations apparently may benefit, this may not be so. Under present globalization of the world, harm to one part of the globe will affect the other parts. There are complex interrelationships between the nations of the world. Present civilization and modernization cannot be sustained without international cooperation. Hence, the effort to combat change must be made worldwide. No individual nation has either the political mandate or economic power to combat climatic change alone. It would be wise to cut the emissions so that the greenhouse changes develop as slowly as possible and ecosystems can adapt gradually to them. People from all sectors, all groups, all nations may need to develop a combined and continuous effort to manipulate the situation. Small and poor countries like Bangladesh, though having little influence on global climate change, would be most harmed; therefore, they should take part in the campaign and should seek cooperation from the international community to doing so.

Reference

Anon (1985) Executive summary. In: Strain BR, Cure JD (eds) Direct effects of increasing carbon dioxide on vegetation. US Department of Energy, Washington, DC, pp XVII–XXV